International Conference on Planetary Boundary Layer and Climate Change 2009

IOP Conference Series: Earth and Environmental Science
Volume 13

Cape Town, Africa
26 - 28 October 2009

ISBN: 978-1-5108-0211-7
ISSN: 1755-1307

Printed from e-media with permission by:

Curran Associates, Inc.
57 Morehouse Lane
Red Hook, NY 12571

proceedings.com

Some format issues inherent in the e-media version may also appear in this print version.

Printed by Curran Associates, Inc. (2015)

Institute of Physics
Dirac House, Temple Back
Bristol BS1 6BE UK

Phone: 44 1 17 929 7481
Fax: 44 1 17 920 0979

techtracking@iop.org

Additional copies of this publication are available from:

Curran Associates, Inc.
57 Morehouse Lane
Red Hook, NY 12571 USA
Phone: 845-758-0400
Fax: 845-758-2634
Email: curran@proceedings.com
Web: www.proceedings.com

Table of contents

Volume 13

International Conference on Planetary Boundary Layer and Climate Change
26–28 October 2009, Cape Town, South Africa

Accepted papers received: 3 August 2010
Published online: 7 September 2010

Papers

IOP Conf. Series: Earth and Environmental Science **13** (2010) 012001 doi:10.1088/1755-1315/13/1/012001

Global Causes and Effects of the Climate Change and Glaciations on the Earth (A Review)

V.M. Kotlyakov

Institute of Geography, Russian Academy of Sciences, Staromonetny St., 29, Moscow 109017, Russia

Email: direct@igras.geonet.ru

Abstract. This review reconciles our knowledge about the past climate fluctuations and natural climate cycles on different time scales. It emphasises the fact that the climate has been and still is highly variable with amplitude of temperature fluctuations exceeding the directly observed range of the last century fluctuations by 1.5 to 2.0 degrees centigrade.

For the last time our planet goes through epoch of the global warming. This period started approximately 150 years ago, having replaced so called "little ice age", i.e. a period of cooling that reached its maximum in the middle of the nineteenth century. For the last century, a rise of the global mean air temperature amounted to a value slightly greater 0.7°C. However, this rise increased during the last 30 years, being especially sharp over the continental regions of Eurasia and North America, and most of all in the Arctic (Fig. 1).

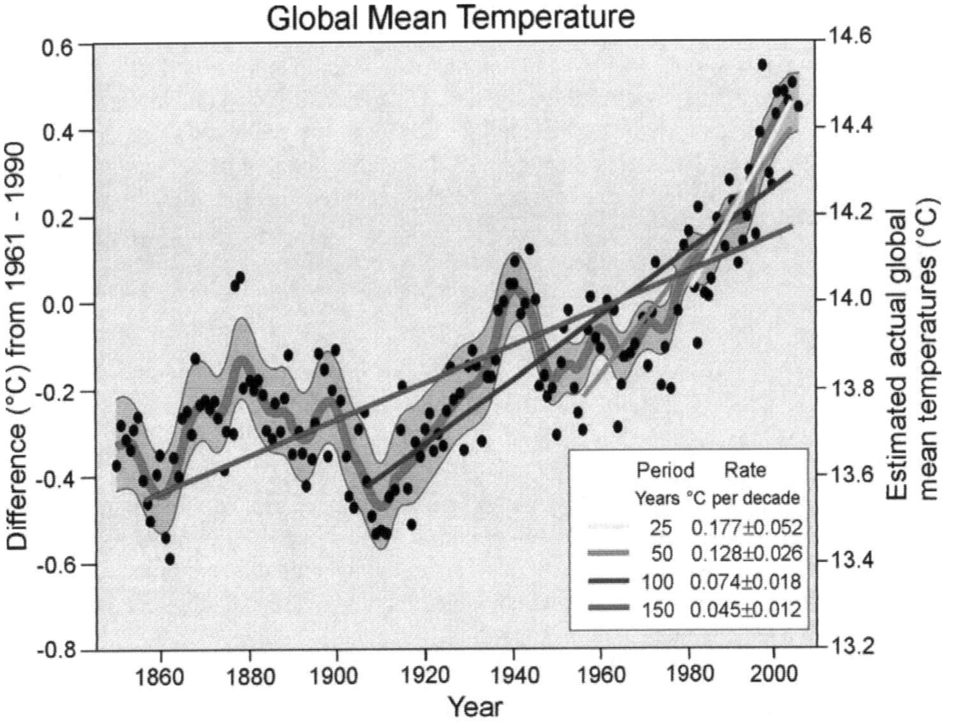

Figure 1. Variation of mean global air temperature for 150 years. Dots are mean annual temperature for each year; the grey band shows deviations of the temperature values. Four straight lines means trends (and rates) of changes of the global temperature for 25, 50, 100, and 150 years.

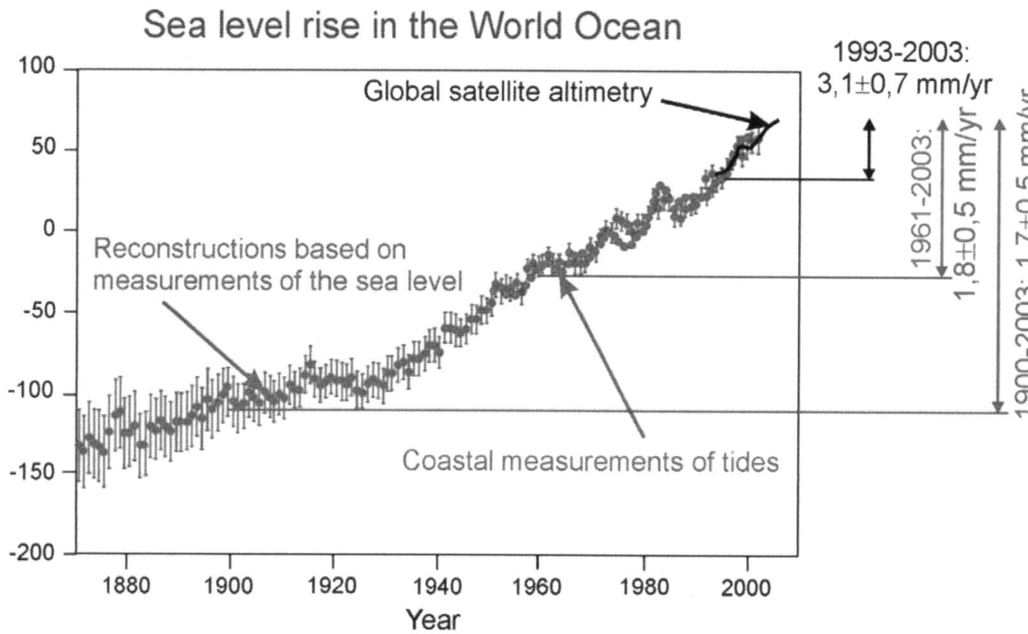

Figure 2. The level rise of the Global Ocean: *1* – reconstructions based on measurements of the sea level (1900–2003); *2* – coastal measurements of tides (1961–2003); *3* – the global satellite altimetry (1993–2003).

At the same time, the sea level rise was recorded too. Reconstruction of this level position in the late nineteenth and early twentieth centuries together with later coastal measurements and then the satellite altimeter observations pointed to the fact of the sea level rise by 1.7 mm per a year in the twentieth century. Note, that during the last decades this rise increased and reached 3 mm per a year (Fig. 2). Causes for the level rise are obviously connected with the temperature rise that leads to expansion of warming surface layer of the Global Ocean, on one hand, while it is caused by melting of glaciers, on the other hand, and thus increasing the water inflow to the ocean [1].

Explanation for this warming is attributed to the anthropogenic effect on the global climate, and for the most part to influence of the greenhouse gases on the temperature rise. For the last decades, development of models describing interaction between the land, ocean, and the atmosphere became especially popular, and now already tens of models are known. This modeling takes into account many factors that determine heat balance of the planet, and, among other factors, great volcanic eruptions. They are marked in Fig. 3 by vertical lines. Such eruptions result in a short-term drop of the global temperature. Modelers state that, if they take into consideration in their models only natural factors, then the calculation results does not correspond to the real temperature course, as this can be seen in Fig. 3a. However, if the modeling is performed with regard for the anthropogenic effects (Fig. 3b), the calculation results are close to the temperature natural variations.

At the same time, an even simple analysis demonstrates that the global course of the temperature variation is much more complicated if it is compared to, for example, a growth of the global consumption of fuel (Fig. 4). It is evident that we are still not able to understand completely all natural processes, which control functioning of the very complicated and complex climatic system and should be used as the basis for modeling of the past and future climate changes.

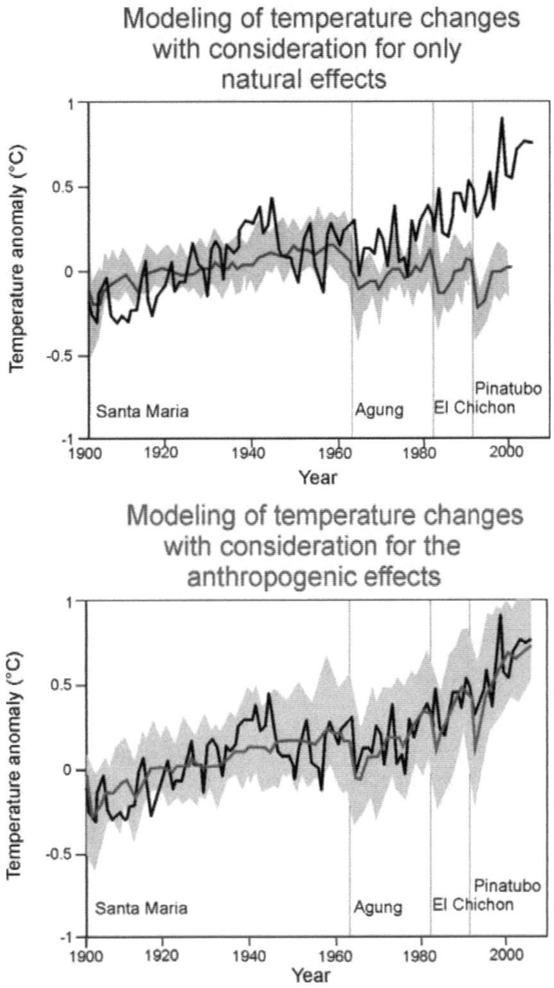

Figure 3. Modeling of the air temperature variation: (*a*) with consideration for only natural fluctuations; (*b*) taking account of the anthropogenic effects. Vertical lines show eruptions of the volcanoes Santa-Maria, Agung, El Chichon, Pinatubo.

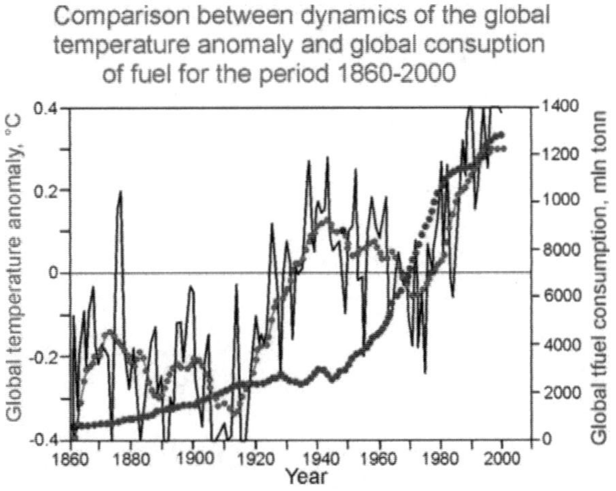

Figure 4. Comparison of dynamics of the global temperature anomalies (*1*) with the global fuel consumption (*2*) for 1860–2000.

When we analyze the global temperature variation, the great role of the solar radiation becomes evident. For the territory to the north of 62° N it is noted that curves of the solar radiation variations and of the mean annual anomalies of the air temperature are similar, however no direct correlation with the carbon dioxide content is revealed. This allows making the inference that a contribution of the Sun's energy into dispersion of the air temperature can amount to 83.5 % while that of the carbon dioxide is only 2.4 %.

Changing of the air temperature is much more drastic in northern latitudes than in the Northern Hemisphere as the whole (Fig. 5a), and this is the general law of the Nature that is still not explained completely. If we average a long-term course of anomalies of mean annual values of the air temperature within the latitude zone to the north from 70° N, we will see a 60-year component in this curve (Fig. 5b).

The current warming exerts very noticeable influence upon the state of perennial ice cover in the Arctic Ocean. Still not long time ago, the heavy ices caused serious complications for navigation along the Northern Sea Route, and the Northwest Passage in the Arctic Archipelago was actually impassable. Presently the closeness of ice greater 70 per cent remains only in the near-pole area and in the north of the Canadian Arctic Islands. Moreover, the total area of the ice cover continuously shortens for the last twenty years. It is recorded that for the last decade area of perennial ices in the Arctic Regions has shortened by approximately 40 %. According to data obtained by the satellite laser altimeter for October, starting with 2004 the average thickness of the sea ice decreased from 2 to 1.4 *m*, area of them shortened by 26 %, and the volume decreased by 50 % [2].

However, if we make a look at the total area of the ices coverage in seas of the Arctic Ocean, we will see regular deviations of this value from linear trends. Here, in the western sector of the Arctic Regions, where the Gulf Stream influence is strong, these fluctuations are more drastic as compared to the eastern sector. These variations are smoother in areas of the Laptev, East Siberian, and Chukchii Seas.

When comparing the ice state in the Arctic Ocean during the first and the second half of the twentieth century (Fig. 6), one can notice essential differences in the above-mentioned sectors. In the first half of the century, area of the sea ices decreased quite quickly in the western seas while in the eastern part it, on the contrary, even slightly increased, whereas in the second half, the same tendency, that is shortening of the sea area, is characteristic for both parts of the Arctic Ocean. In the western seas, the ice area decreased much slower than this took place at the beginning of the century.

Main features of the climate changes are related to fluctuations of the atmosphere general circulation, which mainly depend on a state of the circumpolar vortex. During periods of warming this vortex deepens that results in weakening of the arctic anticyclone and strengthening of the zonal flow in the atmosphere of middle latitudes. Opposite changes are noted during the climate cooling. These features of the atmosphere circulation are explained by variations of the solar activity and the "Sun's dissymmetry" that can change with a period of about sixty years within a range of 2.5% and exert greater influence in the high-latitude regions. Auto-oscillations in the system "atmosphere–ice–ocean" together with influence of aerosols and greenhouse gases of anthropogenic origin also play a certain role in the climate change; however, the decisive factors are natural.

Nowadays glaciers and ice domes on the high-latitude Arctic archipelagos occupy area of almost 250 000 km^2, but the main body of the continental ice is located in Greenland, where the Greenland Ice Sheet and adjacent smaller glaciers and ice domes cover the area more 2 000 000 km^2. During epochs of warming, the glaciers do certainly melt, and their mass decreases that leads to a rise of the Global Ocean level.

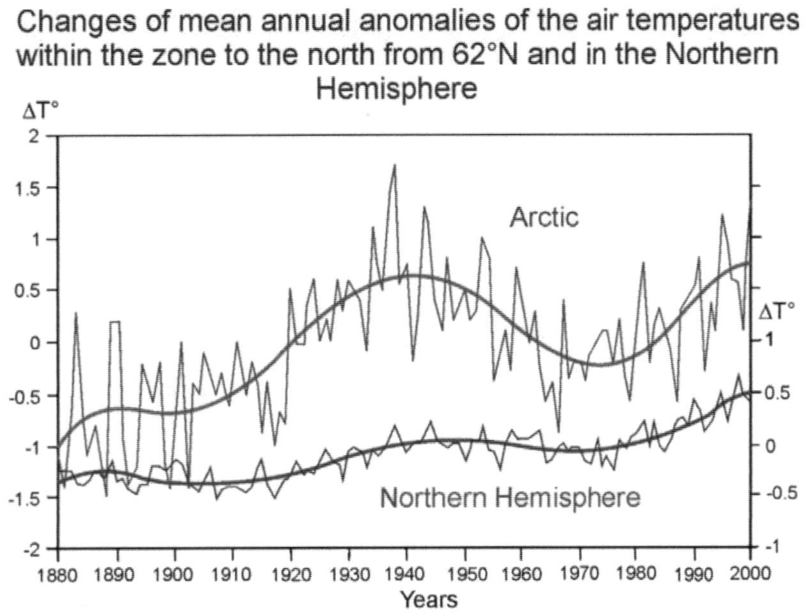

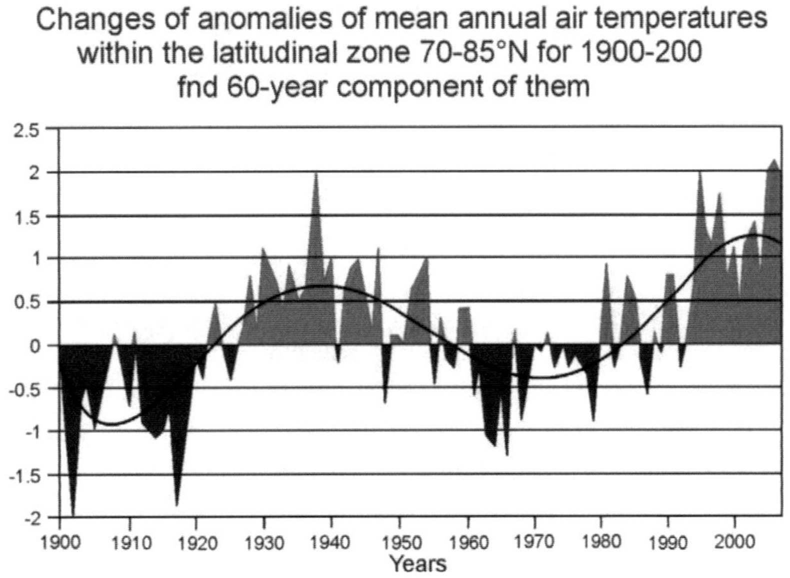

Figure 5. Variation of mean annual anomalies of the air temperature within the zone to the north of 62° N and in the Northern Hemisphere (*a*); anomalies of mean annual air temperature within the latitude zone 70–85° N for 1900–2007 and the 60-year component of them (*b*).

Data obtained by the laser altimeter allow calculating the mass balance of the Greenland Ice Sheet. It is clear that sign and magnitude of this balance depend on an altitude of the ice sheet. Most of the Greenland territory has the positive surface mass balance, and for the whole Greenland, this balance is positive and equal to approximately 170 $km^3/year$. However, considerable ice portions calve from the Greenland Ice Sheet as icebergs, so the total mass balance of this ice sheet is negative and amounts to –200 $km^3/year$. For the last years, these estimates become more and more negative. It means that the current degradation of the Greenland Ice Sheet contribute significantly to the level rise of the Global Ocean.

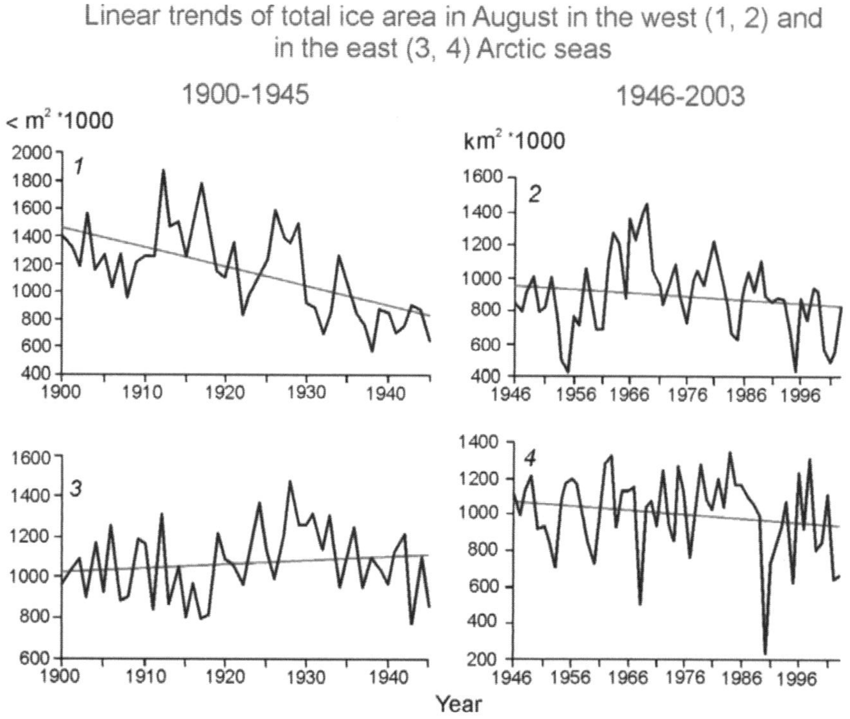

Figure 6. Linear trends of the total area of ice in the western (*1*, *2*) and the eastern (*3*, *4*) Arctic seas for 1900–1945 and 1945–2003.

Velocities of movement of the outlet Greenland glaciers did significantly increase and, respectively, the iceberg calving intensified in recent years. The total negative balance is precisely the result of these processes, and not of intensification of the surface ablation. Different explanations can be proposed as the causes of acceleration of the ice movement. For example, the observed intensification of the surface melting results in that a larger volume of water comes to a bed and, thus, promotes the ice sliding. It is also possible that appearance of warm seawaters in mouths of the outlets glaciers has similar effect.

During the period from 1961 to 2005 average rate of the ice volume change in the Arctic glaciers amounted to –66.1 $km^3/year$ that was equivalent to the ocean level rise by 0.18 *mm/year*. 70 % of the net ice losses in the Arctic regions (without the Greenland Ice Sheet) are connected with a surface loss of the mass (climatic losses), and 30% of them are the ice discharge into the seas (dynamic losses).

By this means, the mass balance of the Arctic continental ice is now negative as a whole. It is difficult now to make a conclusion on how much the revealed balance, dynamic changes of the mass balance of the polar ice sheets are typical, and long lasting since the period of instrumental observations of them cover only a few years. Nevertheless, there are more and more evidences for that the ice sheets are dynamically much more variable than considered before.

Quite another situation takes place in the Antarctica. Comparison between values of the mass balance of the Antarctic Ice Sheet in the middle and at the end of the last century demonstrates increasing activity: both, income and loss of the ice mass here increased. Nevertheless, the total balance remained positive, i.e. for the last fifty years the ice mass continued to grow. It does clearly retard the level rise of the Global Ocean.

IOP Conf. Series: Earth and Environmental Science **13** (2010) 012001 doi:10.1088/1755-1315/13/1/012001

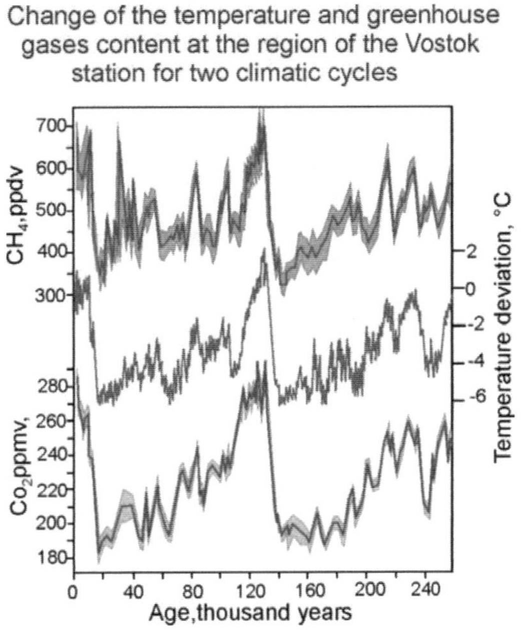

Figure 7. Variation of the temperature and the greenhouse gas content from data of the ice cores from the Vostok deep borehole for two climatic cycles.

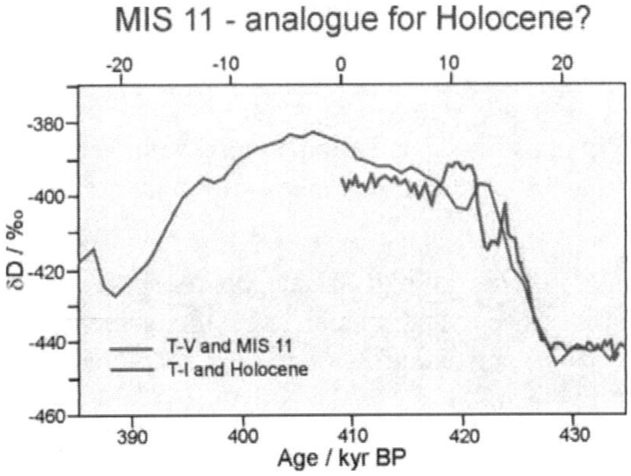

Figure 8. Comparison between the temperature courses (from data on deuterium in the ice samples from glacier boreholes) in the Holocene and in the Interglacial that happened about 400 000 years ago (the sea stage MIS-11).

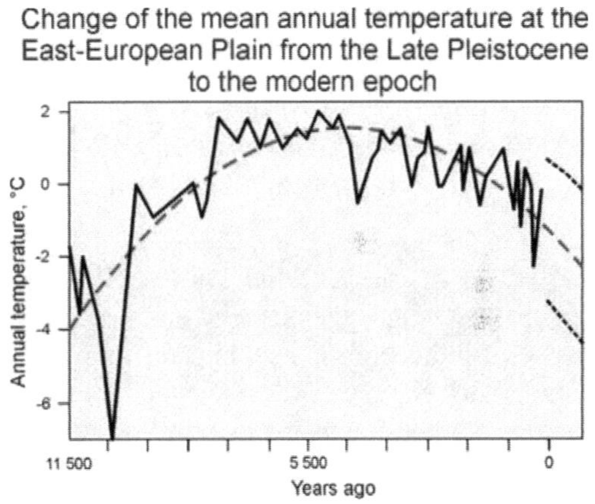

Figure 9. Tendency of the mean annual temperature variation (from the paleo-eographic data) on the East European Plain from the Late Pleistocene to the present time.

Now, let us draw our attention to how the modern climate is related to a climate of the last geological epoch, i.e. the Late Pleistocene and the Holocene. It may be analyzed using data obtained at the Russian station Vostok, located at a height 3.5 *km* in the central part of the Eastern Antarctica. Here, at the Pole of Cold, in 1970s the Soviet Antarctic Expedition started drilling a deep glacial borehole. It reaches now a depth of about 3650 *m*, and for many years, this remains the world record of similar works. Examining of the ice cores, taken at different depths, makes it possible to study the relation between isotopes of oxygen and hydrogen so that to judge from these relations on deviations of the past temperatures from the present values. At the same time, analysis of gases, contained in the ancient ice, allows getting information on a content of the greenhouse gases (carbon dioxide and methane) in the past atmospheres.

As seen in Fig. 7, over the course of two climatic cycles, covering almost 250 000 years, variations of the temperature and the greenhouse gas content were in general going on in parallel: in cold epochs, concentrations of carbon dioxide and methane decreased, while, on the contrary, they increased during the warm periods [3].

Later on, the data for four climatic cycles, covering almost 420 000 years, have been obtained from the Vostok borehole. During this period, we have again found the parallel courses of the air temperature and the carbon dioxide together with similar variation of the Global Ocean level that indicates to corresponding growth and degradation of the ice sheets on the Earth. Finally, similar graph has been recently constructed by European scientists for the Antarctica for the last 750 000 years. It has been also possible to find out that three earlier Interglacials, preceding to the Holocene, are much warmer than the Holocene, i.e. the present-day temperature is still by 1.5–2°C lower than that in earlier periods. It means that, despite a possible anthropogenic impact, the temperature fluctuations on the planet still do not exceed limits of the natural changes characteristic for the whole last geological epoch.

It is pertinent here to note that these data do not allow a general conclusion about what is the first, i.e. a greenhouse growth precedes the warming, or *vice versa*, a warming leads to increase of the greenhouse concentration in the atmosphere. Obviously, there were epochs on the planet, especially during periods of deglaciation, when the warming occurred initially that resulted in growth of the CO_2 and methane concentration in the environment.

One can also see in Fig. 7 that, overall, the Holocene bears no resemblance to previous Interglacials since it is more long and not as much warm as the earlier periods. More likely the Holocene is similar to more ancient Interglacial Epoch, happening on the Earth in about 400 000 years ago (Fig. 8). Changes in natural conditions were also going in similar way on the East European Plain, although they were studied by means of other, i.e. paleogeographic, methods. It follows from these data that changes of the climate and the environment took place at the global scale.

All presented above testified that changes of the global climate proceeded in twentieth century mainly under the influence of natural causes, and changes in the greenhouse gas concentrations in the atmosphere were not always consistent with the climate changes. The climate changes are polycyclic, and the periods of its basic cycles within a century are approximately ten, twenty, and sixty years. It should note here also the main limitations of joined models of the atmosphere and ocean that are used for analysis and prediction of the climate changes. For verification of a response to the anthropogenic impact, the modelers use the data, which characterize mainly natural variations for a short period, and these variations are not directly related to influence of the greenhouse gases. Because of that, scenarios of the temperature changes in twenty first century differ by almost an order.

If now we again look at the Holocene, we see that the warmest peak had been already passed, and this was so-called "climatic optimum of the Holocene" occurred in 5-6 thousand years ago (Fig. 9). After that, the global temperature continues to drop, and the present epoch is situated at the descending section of the curve. Of course, the actual course of the temperature is a combination of cycles having very different amplitudes and duration. For example, during the last millennium the warmest previous epoch falls on ten–eleventh centuries A.D. when Vikings sailed far to North and had discovered the Greenland. Later on, the 16–17th centuries was the cold epoch, well pronounced in history of the Europe and clearly defined in the Russian Chronicles.

Thus, one of fundamental laws of the Nature is its cyclic development. Natural cycles have different characters and periods – from seasonal to 100 000 years. Causes of them are also different and still not completely understood. That is why the climatic models, which are very popular now, demonstrate significant scatter between them, so the future climate is predicted with a low probability.

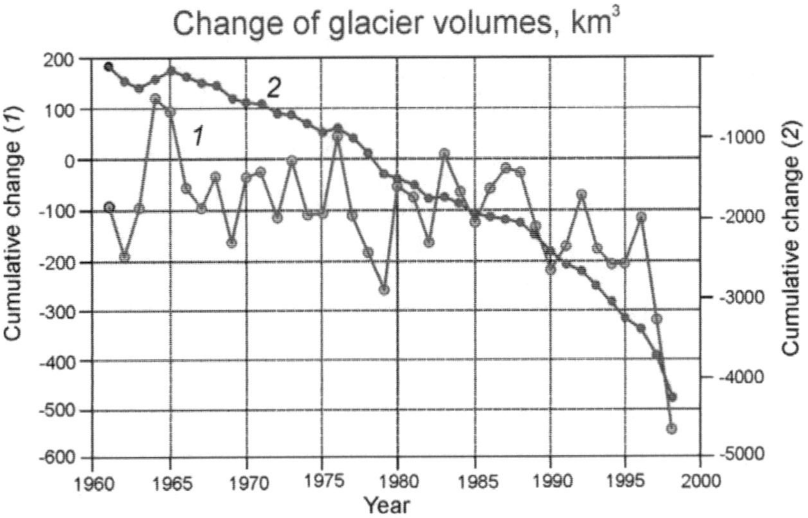

Figure 10. Annual and cumulative change of volumes of glaciers from 1960 until 2000.

It is clear now only that the present epoch is characterized by the global warming that is drastically reflected in a state of glaciers (Fig. 10) and leads to a retreat of them. However, this process happened on the Earth many times and then gave way to colder periods, as it took place quite recently in 1960–1970s. There are no reasons to believe that the current warming will last infinitely long and become more intense. We have no good and thorough scientific basis for such statement.

All contemporary way of the people's life on the Earth (infrastructure, economics, and agriculture) is such that any climatic changes (global or regional warming, cooling, moistening, aridization, etc.) may become unfavorable for the life. The reason is that this way of life was established and rapidly developed during very short period of time that was the last century. It means that the present-day life is adapted to very narrow geographical limits.

Of course, the human society should organize the life, including the economic plans, with consideration for the current climate warming. Nevertheless, at the same time, the humankind must be prepared also to another climate conditions because the Earth's climatic system continues to exist in accordance with its own laws. Fortunately, the anthropogenic impact is still not able to destroy the Nature's laws.

References

[1] Meier M.F., Dyurgerov M.B., Rick U.K., O'Neel S., Pfeffer W.T., Anderson R.S., Anderson S.P., Glazovsky A.F. 2007 Glaciers dominate eustatic sea-level rise in the 21st century, *Science*, **317**, 1064–1067.

[2] Kwok R., Rothrock D.A. 2009 Decline in Arctic sea ice thickness from submarine and ICESat records: 1958-2008, *Geophys. Res. Lett.*, **36**, L15501, doi:10.109/2009GL039035.

[3] Jouzel J., Barkov N.I., Barnola J.M., Bender M., Chappellaz J., Genthon C., Kotlyakov V.M., Lipenkov V., Lorius C., Petit J.R., Raynaud D., Raisbeck G., Ritz C., Sovers T., Syievenard M., Yiou F., Yiou P. 1993 Extending the Vostok ice-core record of palaoclimate to the penultimate glacial period, *Nature*, **364**, 407–412.

IOP Conf. Series: Earth and Environmental Science **13** (2010) 012002 doi:10.1088/1755-1315/13/1/012002

Cryosphere and Climate

T.E. Khromova
Institute of Geography, Russian Academy of Sciences, Staromonetny St., 29, Moscow 109017, Russia

Email: khromova@igras.geonet.ru

Abstract. Analysis of current and historical data demonstrates that the cryosphere component are very sensitive to the climate change thus proving its role as the climate indicator. In the process of this a pattern of the changes clearly reflect regional climate characteristics. International cryospheric community works very hard for efficient organization of observations of the cryosphere components and specific features of its development against background of changing climate. Geophysical, geochemical, space and other methods are used for solution of this problem.

1. Introduction

Cryosphere is one of the Earths spheres between the atmosphere, hydrosphere and lithosphere, where negative or zero temperatures prevail and water is present in solid or super-cooled state [1]. The basic components of the cryosphere are snow cover, glaciers, ice sheets, sea ices and icebergs, permafrost, river and lake ice etc.

Like at the present time, the Earth history in the Pleistocene is intimately linked with the natural processes running within the cryosphere itself and in its interaction with the atmosphere, hydrosphere and the land surface. However, many processes of such interaction as well as consequences of them are still not sufficiently studied. Note, that the natural ices themselves are rather changeable, and under conditions of the global warming serious changes proceed in the cryosphere whose results are of the global scale (sea level rise, permafrost degradation, etc.). Prediction of future state of the environment, in particular functioning of the climate system, can not be understood without thorough analysis of the cryospheric processes.

2. Components of the cryosphere

The most widespread component of the cryosphere is the snow cover. At the average, it covers almost 26 million km^2, for the most part in the Northern Hemisphere. Mean area of the snow cover ranges here from 46.5 million km^2 in January down to 3.8 million km^2 in August [2].

Sea ice covers the most part of polar oceans. In the Southern Hemisphere area of seasonal ice cover varies from 3–4 million km^2 in February down to 17–20 million km^2 in September, while in the Northern Hemisphere from 7–9 million km^2 in September to 14–16 million km^2 in March [3].

Ice on rivers and lakes is formed because of the seasonal cooling. Intensity of processes of the ice cover formation and destruction is highly variable and depends on the weather conditions. Long series of observations of the lake ice cover can be used as a climate indicator.

Like snow, seasonally frozen ground cover vast expanses. Their depth and spread depend on the air temperature, snow depth, vegetation, and humidity of soil. Perennially frozen ground or permafrost may occur where mean annual air temperature drops below –1 °C, and is generally continuous at temperatures below –7 °C. Permafrost occupies approximately 24.5 % of the Northern Hemisphere, and its maximum is located between 60 and 68° N. Maximal thickness of the permafrost reaches 600 m in north-east of Siberia and in the Alaska.

About 77% of total global ice is concentrated in ice sheets. It corresponds to a layer of the World Ocean of about 71 m. 90% of this quantity is accumulated in the Antarctica, and 10% is in Greenland. Other glaciers contain less 0.5 %. Antarctic and Greenland ice sheets have essential effect on the atmosphere circulation, cyclonic system, and the global energy budget. Big areas of the West Antarctic Ice Sheet are located below the sea level, but the main mass of the Antarctic Ice Sheet lies on the bedrock. In the central part of the sheet ice melts on the bed and this forms sub-glacial lakes. The largest of them is situated near the Russian station Vostok. Ice shelves and outlet glaciers do constantly produce icebergs: every year about 2072±304 km^3 of ice calves in the Antarctica, and about 235±33 km^3 in Greenland [4].

Components of the cryosphere are present in all latitudes of the globe, including equator. In January snow cover and sea ice occupy in average 23 % of the Northern Hemisphere total area. Changes of sizes and characteristics of the cryosphere components, especially of global snow and ice cover, can have grave consequences for the global climate system. Changes in the cryosphere will have also certain social-economic consequences, especially in cold regions. Elements of infrastructure (buildings, electric power lines, roads) are in the zone of risk, hydraulic power engineering depends on a regime of snow and ice melting, and the winter tourism is directly connected with amount of snowfall. Constructions and transport can be prone to impact of glaze and hoarfrost phenomena, and summer melting of glaciers and ice in mountains is essential for water supply and irrigation. The lowlands can be flooded due to the sea level rise because of melting of glaciers and ice sheets.

3. The cryosphere role in the climate system

Observation and monitoring of all components of the cryosphere is of extreme importance for understanding of processes in the cryosphere and the earth climate system. Owing to their unique physical properties, snow and ice play essential role in the climatic processes. They have relatively high albedo and reflect larger amount of short-wave solar radiation than surface of a land or water (Fig. 1). In many regions, the snow cover and its melting determine regime of water supply, hydrological cycle, filling of water basins and runoff to the World Ocean. Perennial and seasonal frozen rocks change flows of water, energy and gases between surface and the atmosphere and have strong influence upon state of land surface, hydrology and vegetation.

Processes of freezing and melting exert serious impact upon water, energy and chemical cycles. Glaciers and ice sheets conserve water during hundreds and thousands of years. If the whole ice on the Earth will melt, then the World Ocean level can rise by 60–70 m (Fig. 2). Formation and destruction of sea ice, processes on lower surface of ice shelves and around icebergs redistribute fresh and salt water. So, one can say that the transformation of water masses control the global thermo-haline circulation.

The cryosphere is an integral part of the global climate system with many direct links and feedbacks which are formed through effect on near-surface fluxes of energy and moisture, clouds, precipitation, atmospheric and oceanic circulation (Fig. 3). Main factors of this interaction are high albedo of the snow and ice surfaces, latent heat, involved into phase transition ice/water, isolating effect of snow cover on ground surface and of ice on water surface, delays in every-year energy and water cycles caused by seasonal snow and ice cover, greenhouse gases included in the frozen rocks.

Under conditions of the changing climate, the feedback and intensification of the climate signal through interaction with the cryosphere plays have a special role. The feedback being caused due to high albedo of snow and ice as well as the greenhouse gas emissions in areas of the permafrost existence can essentially enhance the warming. Many global climatic models predict the greatest warming in the Arctic region due to changes of high albedo in the past time. The enhancement of melting of the sea ice in the Arctic and of ice sheet in Greenland and increasing of the north river runoff result in desalination of surface waters in the North Atlantic. This can in its turn cause moderation of meridional circulation in the World Ocean including weakening of the North-Atlantic current, which determines relatively mild climate in the north-west of Europe. These processes can have serious impacts for future of both the European and the global climates.

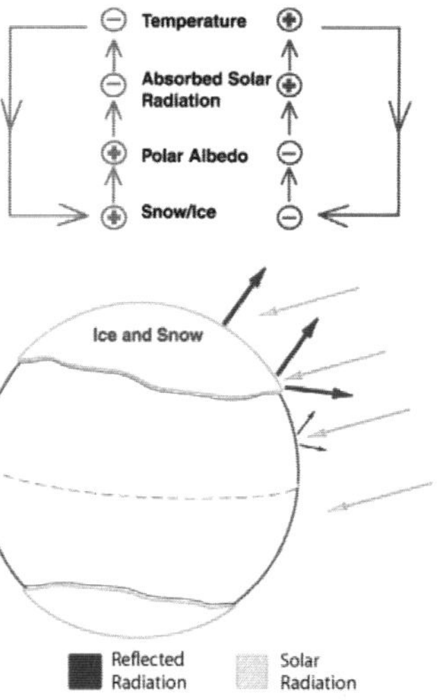

Figure 1. Schematic representation of the snow–ice/albedo feedback.

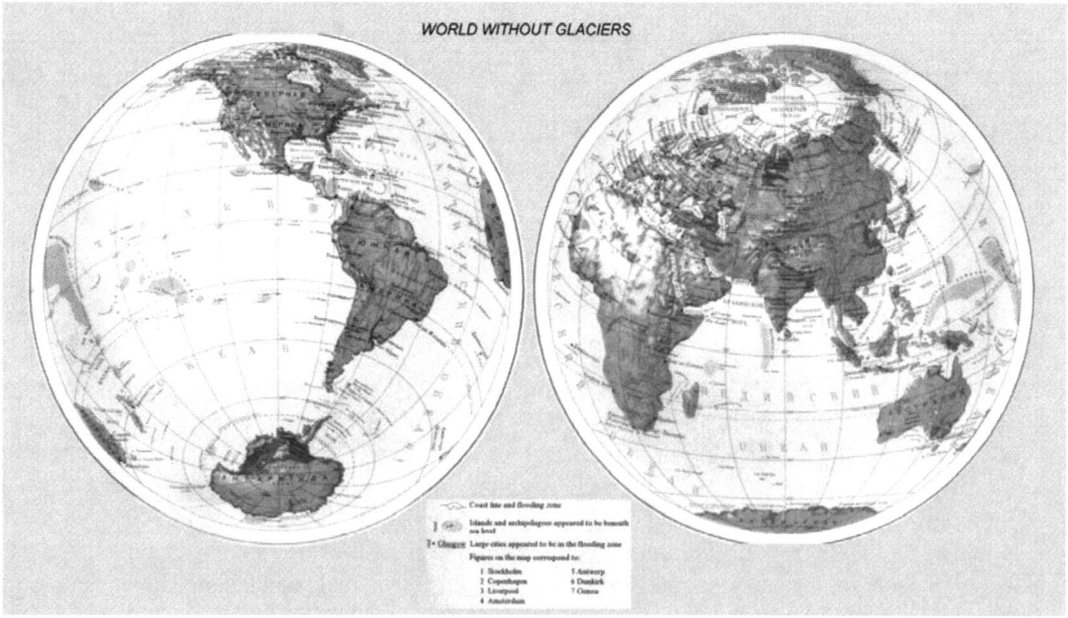

Figure 2. World without glaciers.

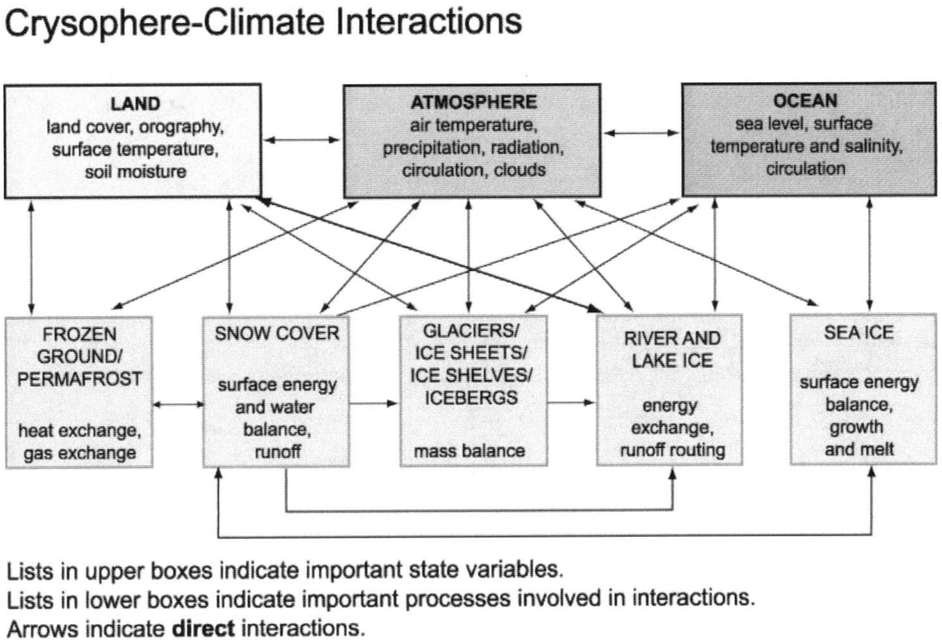

Figure 3. Schematic representation of the cryosphere–climate interactions. After G. Flato (AES, Climate Research Branch).

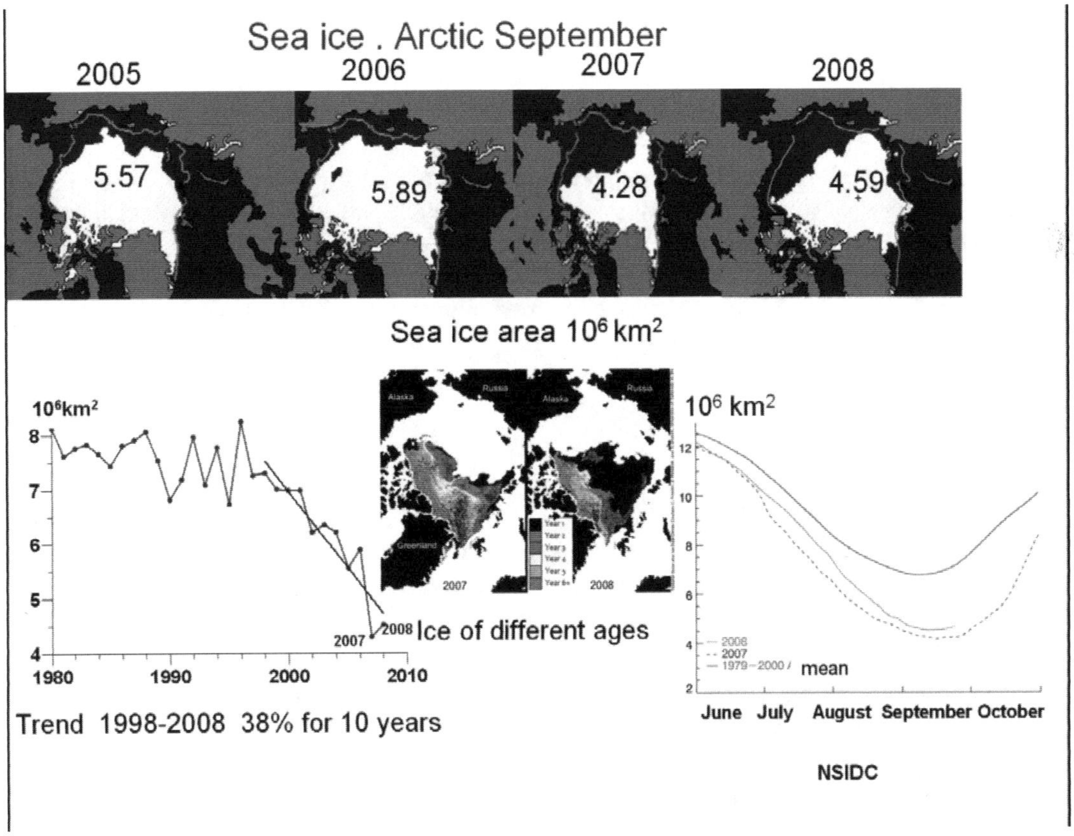

Figure 4. Changes of the sea ice area in Arctic taken from publications [3, 5, 6].

4. The present state of main cryosphere components

Many processes of interaction between the cryosphere and the climate system have not been adequately investigated. For the most part the cryosphere is not easily accessible for direct measurement, so the studying of it requires large material expenditures.

Sea ices. The most important instruments for the sea ice monitoring are methods of remote sensing, in particular the satellite surveys in different ranges and the laser altimetry. They make possible to monitor areas and thicknesses of the sea ices. Available models are still not able to predict adequately changes of their main characteristics. Starting with 1970s, the temperature measurements in the Arctic show steady trend of warming. In parallel with this, essential early fluctuations relative to average values take place. During the very last years (2007–2009) a certain cooling is noted in the Arctic that has already caused increase of the sea ice areas (Fig. 4).

Permafrost. The current air temperature rise causes degradation of permafrost. Further warming can lead to increasing of the thawing layer, rising of the rock temperatures, and as a result to a drastic decrease of the permafrost layer. The permafrost is very sensitive indicator of climatic variations, so the information about its state and changes is included into integral databases, in particular for the North Eurasia. These databases are used as bases for modeling of spatial spreading of the permafrost and dynamics of its active layer (Fig. 5). Even though investigations are wide and active, there are some uncertainties in estimating of processes running within the permafrost zone. Measurements of the soil temperature in some Sub-Arctic regions do not show any definite trend, and this does not allow receiving evident quantitative estimates of volumes of the natural carbon cycle.

Snow cover. Main source of information on the snow cover are data of remote sensing and observations at meteorological stations. There are several archives of the remote sensing data, used as basis for development of systems for map-making the snow cover distribution and thicknesses. The NOAA Interactive Multisensor Snow and Ice mapping system is actually more than 35-year period of constructing maps of snow and ice distribution in the Northern Hemisphere. Space images in visible range make it possible to use the interactive analysis. Gradations of spatial and time resolution include weekly data with accuracy of 180 km (1972–1998), daily data with accuracy of 24 km (1998–2004), and daily data with accuracy of 4 km (since 2004 till present time). MODIS snow map (NASA) are global daily maps with resolution 0.5 and 5 km, available since 2000. One more source of the information is presented by the satellite Terra and Aqua. AVHRR/NOAA snow/ice map (NESDIS) provides daily global coverage with resolution of 4 km, starting from 2006. Main shortage of all these data is cloudiness that does not allow analysis of snow cover over approximately 40 % of the globe surface.

Investigations show that in many regions duration of seasonal snow cover occurrence as well as a period of snow melting become shorter. For example, area of seasonal snow cover in Tien Shan has been shortened for 2000/01–2006/07 by 15 %. However, despite variety and abundance of the remote data there are still many problems, which prevent obtaining relevant and qualitative notion of spatial snow cover spreading. Nevertheless, there is still no uniform qualitative instrumentation for mapping of snow cover, and possibilities to estimate snow storage are limited.

There are also many problems with ground measurements of solid precipitation at meteorological stations. Observational networks in mountains and high latitudes is reduced, methods of observations change, that makes difficulties in using of data for past periods and introduces additional errors into date of measurements (underestimating or overestimating of the precipitation sums due to wind transport, losses due to moistening and evaporation, disregard at small values, etc.). As before, there are no observations of precipitation above surface of the World Ocean. All these problems have to be solved including needs of data for numerical modeling.

Glaciers. Main source of information about state of glaciers in the second half of the twentieth century is the World Glacier Inventory. These data are available at site of the World Data Center of Glaciology in Boulder Co, USA. According to data of observations, the glaciations is now in a stage of degradation, but for correct estimation of the glacier melting contribution into the ocean level rise we need more careful monitoring and modeling.

The present-day data on the glaciers come, first, from remote sensing. International projects are organized, such as GLIMS and GobGlacier, which are aimed at studying of ground glaciations by means of the remote sensing. A system of regional centers is created in framework of the GLIMS project. One of those functions in the Institute of Geography, RAS and is responsible for glacial regions of the Northern Eurasia. Original software has been developed in framework of this project that makes possible to decipher glacier boundaries and to save results of deciphering in the standard form intended for storage and analysis in a special data base. Specialists of the Institute of Geography have created vector coverage of contours of more than 15,000 glaciers in the Arctic, Polar Urals, Caucasus, Pamirs, Tien Shan, Altai, Kamchatka, Siberia territories. The data obtained testify to shortening of glaciers in all regions of the investigation.

Investigations carried out in other regions of the world also demonstrate that the glaciations are shortened. The greatest attention is called to Polar Regions where main masses of ice are concentrated. Whereas estimates of the shortening are in good agreement, the estimates of changes of the ice volume are differ dramatically. For example, according to different sources changes of the ice volume in Greenland vary from +75 to –70 km^3 per year (Fig. 6). Significant reduction of the ice area is noted for archipelagos of the Russian Arctic. However, the instrumental observations cover relatively short time, so these data does not allow inferring on the Arctic glaciations behavior in a long-term aspect.

The problem of correct estimation of the ice storage on the planet is now calling. Total inventorying of glaciers is carried out in 11 countries, and some national inventories are not accessible for foreign users. The World Glacier Inventory contains information on almost 37 % of the glaciations, and the data base of the GLIMS project contains information about 39 % of number of glaciers and 27% of an area of the whole glaciations.

5. Co-ordination of the cryosphere studies

Without understanding the cryosphere role in the climate system, it is not possible to model and predict its variability and future changes. Therefore, we have to stimulate and co-ordinate further investigations of the cryosphere components and interaction between the cryosphere and climate. This is the goal of the project "Climate and Cryosphere" (CliC) [9] of International Program of the Climate Investigation. Main tasks of the project are as follows: improving of global and regional climate models of high latitudes; extending of knowledge about feedback within the system "the cryosphere–climate"; understanding of the cryosphere processes forming the basis of the global climatic relations; determination of the pathways of moisture transport that control the mass balance of ice sheets; understanding of drastic climate changes related to impact of the cryosphere on the climate system; clearer view on the role of cryosphere in global water, energy, and carbon cycles. To co-ordinate observations of the cryosphere and to measure its main parameters the cryosphere had been separated in 2004 as independent direction in framework of the Integrated Global Observing Strategy (IGOS). The problem of creating of integrated and coordinated observational system is calling. It should include a system of satellite, aircraft and ground observations, modeling and assimilation of the data as well as a reanalysis, organization of the data and information, storage of historic data and others.

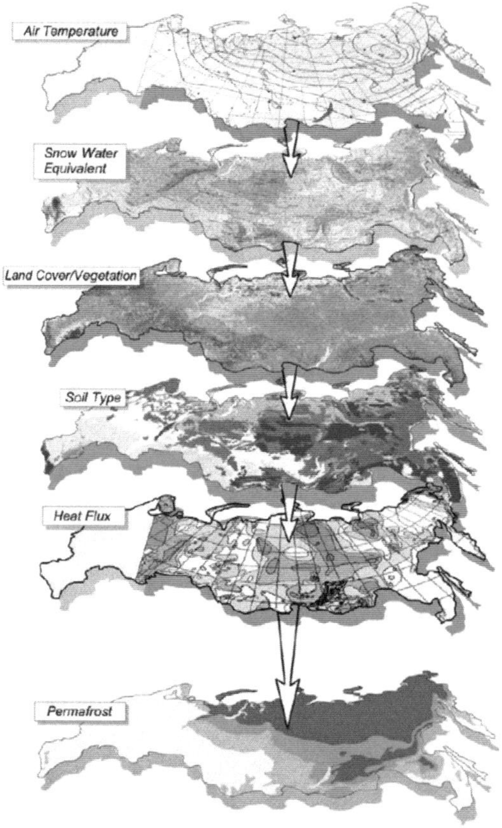

Figure 5. The GIS multi-layered database model for Spatially Distributed Permafrost and Active Layer Dynamics modeling in the Northern Eurasia (after Vladimir Romanovsky, Geophysical Institute, University of Alaska, Fairbanks).

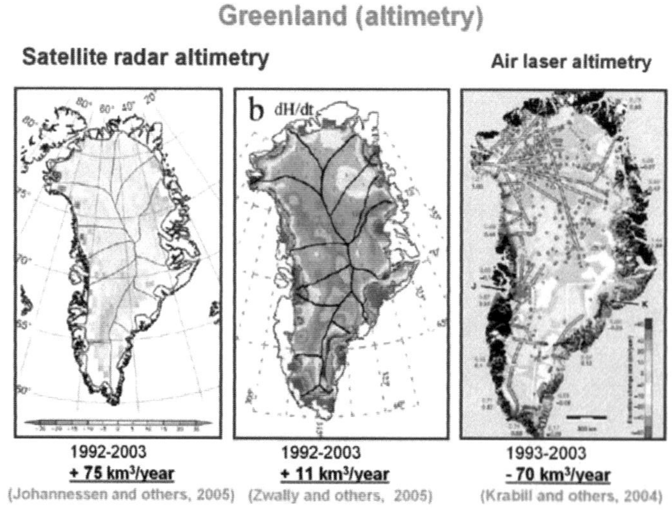

Figure 6. Estimates of the ice storage in Greenland from publications [5, 7, 8].

References

[1] Kotlyakov V.M., Komarova A.I. 2007. Geography. Concepts and Terms. Dictionary in five languages, Moscow, *Nauka*, 859 pp.

[2] Robinson, D.A., K.F. Dewey and R.R. Heim, 1993. Global snow cover monitoring: an update, *Bull. Am. Meteorol. Soc.*, **74**, 1689-1696.

[3] Gloersen, P., et al., 1993. Arctic and Antarctic sea ice, 1978-1987. Satellite passive microwave observations and analysis. *NASA SP-511*, NASA, Washington, DC, 290 pp.

[4] Church, J.A., J.M. Gregory, Ph. Huybrechts, M. Kuhn, C. Lambeck, M.T. Nhuan, D. Qin, and P.L. Woodworth, 2001. Changes in sea level. Chapter 11 in: IPCC Third Scientific Assessment of Climatic Change, J. T. Houghton, Y. Ding, D.J. Griggs, M. Noguer, P. J. van der Linden and D. Xiaosu (Eds.), *Cambridge University Press*, UK, 944 pp.

[5] Johannessen, O.M., Khvorostovsky, K., Miles, M.V., Bobylev, L.P., 2005. Recent Ice-Sheet Growth in the Interior of Greenland, Science, 1013-1016.

[6] Parkinson, C.L., D.J. Cavalieri, P. Gloersen, H.J. Zwally and J.C. Comiso, 1999. Arctic sea ice extents, areas, and trends, 1978-1996. *J. Geophys. Res.*, **104**(C9), 20837-20856.

[7] Krabill, W., W. Abdalati, E. Frederick, S. Manizade, C. Martin, J. Sonntag, R. Swift, R. Thomas,W. Wright, J. Yungel, 2000. Greenland Ice Sheet: High-Elevation Balance and Peripheral Thinning, Science, **289**, 428 – 430.

[8] Zwally, H.J., Giovinetto, M.B., Li, J., Cornejo, H.G., Beckley, M.A., Brenner, A.C., Saba, J.L., Yi, D., 2005. Mass changes of the Greenland and Antarctic ice sheets and shelves and contributions to sea-level rise: 1992-2002, *J. Glaciology*, **51**, 509-527.

[9] CLIC, 2001. Climate and Cryosphere Project Science and Co-Ordination Plan. Version 1. Eds. Allison, I., Barry R.G., Goodison, B.E., WCRP-114 WMO/TD No. **1053**. http://clic.npolar.no/introduction/science_plan.pdf

IOP Conf. Series: Earth and Environmental Science **13** (2010) 012003 doi:10.1088/1755-1315/13/1/012003

A Stable Boundary Layer Perspective on Global Temperature Trends

R.T. McNider, J.R. Christy, A. Biazar

Earth System Science Center, University of Alabama in Huntsville
Huntsville, AL 35899

Email: mcnider@nsstc.uah.edu

Abstract. One of the most significant signals in the thermometer-observed temperature record since 1900 is the decrease in the diurnal temperature range over land, largely due to warming of the minimum temperatures. While some data sets have indicated this asymmetrical warming has been reduced since 1979, regional analyses (e.g. East Africa) indicate that the nocturnal warming continues at a pace greater than daytime temperatures. The cause for this night time warming in the observed temperatures has been attributed to a variety of causes. Climate models have in general not replicated the change in diurnal temperature range well. Here we would like to try to distinguish between warming in the nocturnal boundary layer due to a redistribution of heat and warming due to the accumulation of heat. The temperature at night at shelter height is a result of competition between thermal stability and mechanical shear. If stability wins then turbulence is suppressed and the cooling surface becomes cut-off from the warmer air aloft, which leads to sharp decay in surface air temperature. If shear wins, then turbulence is maintained and warmer air from aloft is continually mixed to the surface, which leads to significantly lower cooling rates and warmer temperatures. This warming occurs due to a redistribution of heat. As will be shown by techniques of nonlinear analysis the winner of the stability and shear contest can be very sensitive to changes in greenhouse gas forcing, surface roughness, cloudiness, and surface heat capacity (including soil moisture). Further, the minimum temperatures measured in the nocturnal boundary layer represent only a very shallow layer of the atmosphere which is usually only a few hundred meters thick. It is likely that the observed warming in minimum temperature, whether caused by additional greenhouse forcing or land use changes or other land surface dynamics, is reflecting a redistribution of heat by turbulence—not an accumulation of heat. Because minimum temperatures in the stable boundary layer are not very robust measures of the heat content in the deep atmosphere and climate models do not predict minimum temperatures well, minimum temperatures should not be used as a surrogate for measures of deep atmosphere global warming.

1. Background

Observations have been a key factor in validation of model projections of heat accumulation in the atmosphere due to atmospheric greenhouse forcing. Surface observations of global temperature have shown overall warming since the middle of the 19th century providing support that heat is accumulating in the atmosphere. There is, however, a curious aspect of the surface data in that there is a strong asymmetry in warming between day and night through most of the temperature record [1]. The most generally used surface data sets (Goddard Institute for Space Studies [2] and East Anglia/Hadley Center [3] only include one temperature for the day, T_{mean}, the daily average of the maximum temperature, T_{max}, and minimum temperature, T_{min}. However, data sets such as the NOAA Global Historical Network [4, 5] which partition the global temperature into minimum and maximum temperatures for a day show that nearly 2/3 of the warming over land has occurred in minimum temperatures (see figure 1). While [6] have shown that the differential warming has decreased significantly since 1979, the number of stations in the GHCN since 1979 has also decreased substantially with very little coverage in the developing world. Analyses of East Africa temperatures [7] in fact show increasing differences in day and night warming since 1979.

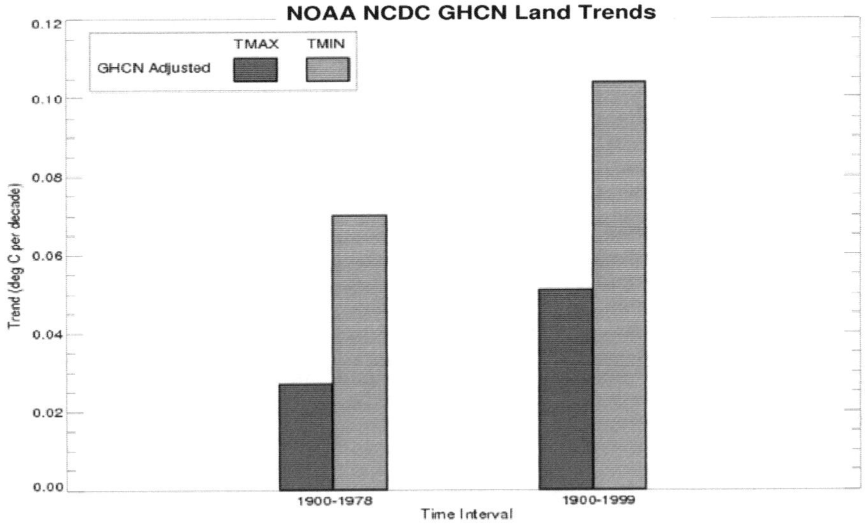

Figure 1. Trends in minimum and maximum temperatures from the NOAA GHCN data set.

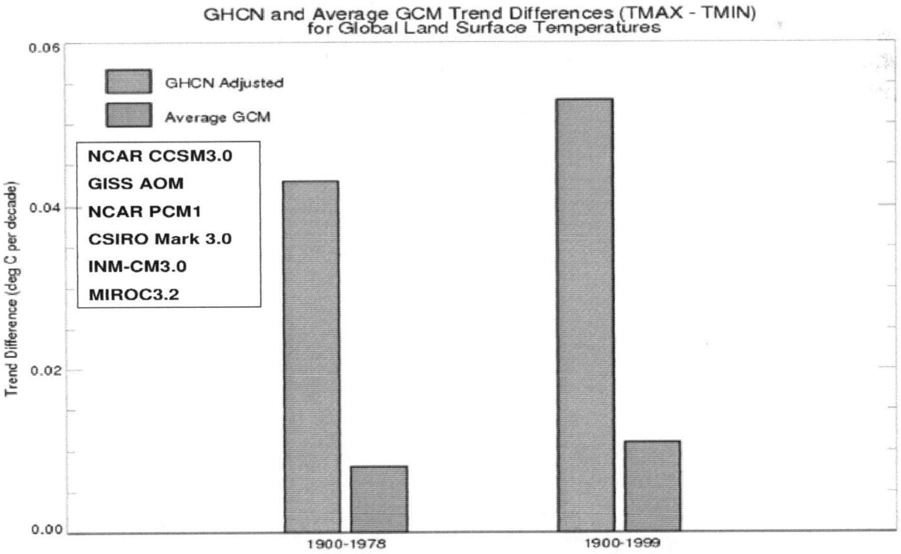

Figure 2. Trends in minimum and maximum temperatures from an average of five global models compared to the GHCN data set.

Despite the fact that this diurnal asymmetry in warming is one of the most significant signals in the historical data, climate models including greenhouse gas forcing have in general not captured the magnitude of this asymmetry. In a sampling of 7 climate models used in a DOE Model inter-comparison study we found the difference in trend between T_{max} and T_{min} to only be 20% of the trend difference in the GHCN data set (see figure 2). This is consistent with the findings of [8] who found little skill in GCM diurnal trends and [9] who found only 33% of the trend difference in the GCM. Because climate models with greenhouse gas forcing have understated the nocturnal warming, investigators have attributed the change to a variety of causes not well represented in climate models such as increased cloud cover, jet contrails, and changes in surface characteristics such as land cover and land use e.g. [10–13].

We have taken an alternative approach and examined the role nonlinear boundary layer dynamics [14–18] might play in the behavior of the nocturnal boundary layer (NBL). Using numerical continuation [19] for a simple two layer model in a paper led by the CMG atmospheric graduate student [20], we found that slight changes in incoming longwave radiation, can destabilize the NBL. This in turn mixes warm air to the surface leading to warming in surface temperature (see figure 3). In an analysis of the dependence of solution on grid spacing the paper further proposed that coarse grid global climate models (GCMs) may not have the ability or the type turbulent closures that can capture this nonlinear behavior. Durre and Wallace [11] had alluded to complexities of the nocturnal boundary layer in impacting the asymmetry.

We believe that additional work is required to understand the sensitivity of the NBL to anthropogenic forcing and to improve the ability of GCMs to simulate NBL processes. While a recent National Academy Report [21] and CCSP [22] report attempted to reconcile differences between surface data trends and deep atmospheric trends (balloon and satellite observations), we believe there is still work to be done in this area especially in regards to NBL processes which were not a focus of the CCSP report [23].

Building mathematical tools that can help understand climate sensitivity is critical especially in the NBL where most of the observed warming has occurred. The following describes recent work that examines the role of stable boundary layer dynamics in the interpretation of climate trends. Additionally, because of issues with land use effects on trends and our current inability to model the stable boundary layer in coarse grid models, we propose that minimum temperatures should not be used to infer the accumulation of heat in the deep atmosphere.

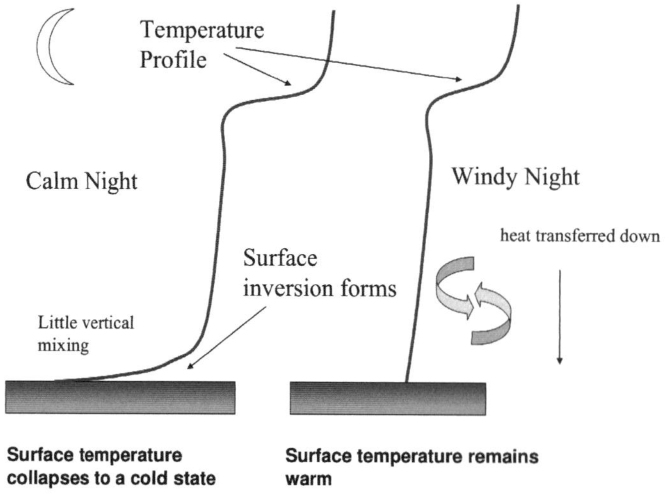

Figure 3. Schematic of transition of stable boundary layer from calm cold solution to warm windy solution.

2. Analysis of Trends in East Africa

As noted above recent analyses of the GHCN data have indicated that the difference in minimum and maximum trends have been reduced since 1979 [6]. However, in examining this data set there have been a substantial reduction in the number of stations in the data set (figure 4) and the distribution of the data set is decidedly concentrated in the developed world especially the U.S. (figure 5). There are few sites in the developing world. Here we examine the temperature trends in areas of East Africa that have had substantial population growth and land development. Christy et al. [7] report on a detailed analysis of trends and possible causes in East Africa.

Most of the global climate data sets only have a few sites in a region. As an example for the East Africa study the standard data sets such as GISS [2], CRUT [3] and GHCN [4] had approximately 15-25 stations in Kenya and Tanzania. In contrast, *Christy et al., 2009,* employed 118 stations. Figure 6 shows the different trends in the data sets for East Africa. Note that the normal global data sets with a small number of stations have a different and larger trend than the full set of stations. The major datasets used only a handful of stations (1 to 10) for the East Africa grid. For that same grid we used 46 stations. This may be due to the standard data sets using sites that have had more land development and population growth.

The difference between minimum temperatures (T_{Max}) and maximum temperature (T_{Min}) trends is striking in these data sets. The large asymmetries in warming rates in T_{Max} and T_{Min} in the 1979-2004 in the East Africa data set [7] are thus at odds with the global findings of [6] and in fact the asymmetries are more like the global asymmetries found in the 1900-1979 period reported by [1]. There is a concern about the recent trends reported by [6] in that the number of stations in the adjusted analysis drops from nearly 3000 stations in 1970 to less than 1500 in 2004 (figure 3). Further, the locations of stations is not well distributed around the globe with few stations in the developing world in South America, India and Africa that are similar socio-economically to East Africa (figure 4). See [7] for additional details.

3. Role of the Stable Boundary Layer in Minimum Temperature Trends

Investigators have attributed the differential trends in maximum and minimum temperatures to a variety of causes including increases in atmospheric water vapor, cloud cover, jet contrails, and changes in surface characteristics such as land cover and land use [10, 11]. Few, however, have considered the effect on the minimum nocturnal temperature of the highly non-linear stable nocturnal boundary layer (SNBL). For a constant mean boundary-layer heat input, Pielke and Matsui [24] demonstrated that measured "surface" nocturnal temperatures are highly dependent upon wind speed and the height above ground that the measurements are made. However, they used a static model that did not include forcing-induced changes in the boundary-layer structure. Durre and Wallace [11] noted that an increase in greenhouse gases could potentially affect weak longwave radiation and thus alter SNBL processes, but they did not discuss the processes.

The SNBL is characterized by competition between mechanical generation of turbulence from wind shear and the suppression of turbulence by thermodynamic stability [25]. Using a single-column model truncated to two layers and employing bifurcation techniques, McNider et al. [26] showed that parameter spaces exist in which the SNBL is highly sensitive to slight changes in wind speed and surface roughness which lead to dramatic changes in nocturnal surface temperatures and multiple solutions. Van de Wiel et al. [14, 15] also carried out a dynamical analysis; however, their model did not allow interaction with the outer atmosphere, so they did not have multiple solutions.

The bifurcation techniques developed are used here to explore parameter dependence of land use variables and demonstrate that weak radiative forcing (4-6 W m^{-2}); such as caused by an increase in greenhouse gases or aersosols, can trigger a potentially significant positive feedback in the SNBL. The analysis uses a simplified formulation of the SNBL to explore the sensitivity and role of boundary processes. It does not include the cumulative effect of these processes as would a true climate simulation.

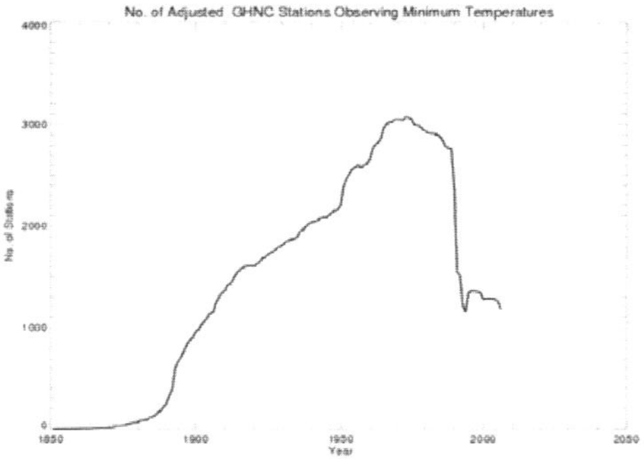

Figure 4. Number of stations included in the GHCN data set.

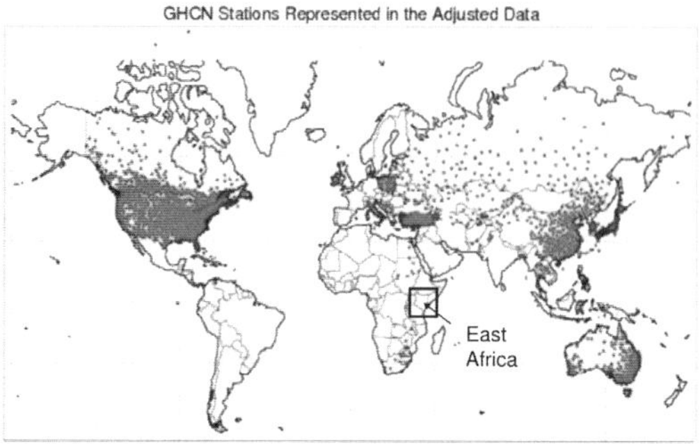

Figure 5. Geographical distribution of GHCN observations.

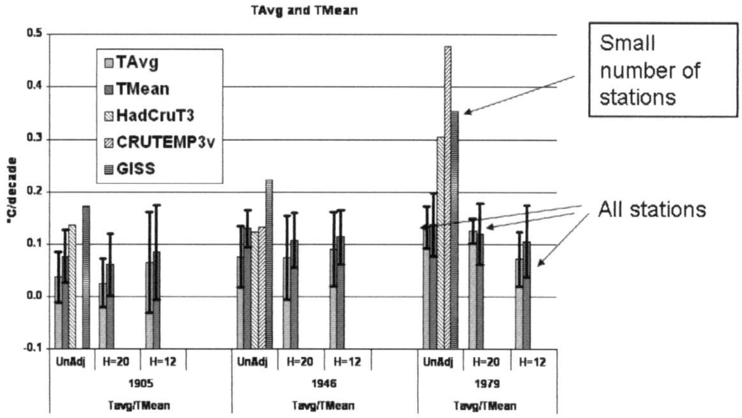

Figure 6. Trends in T_{mean} and T_{avg} in East Africa from [7].

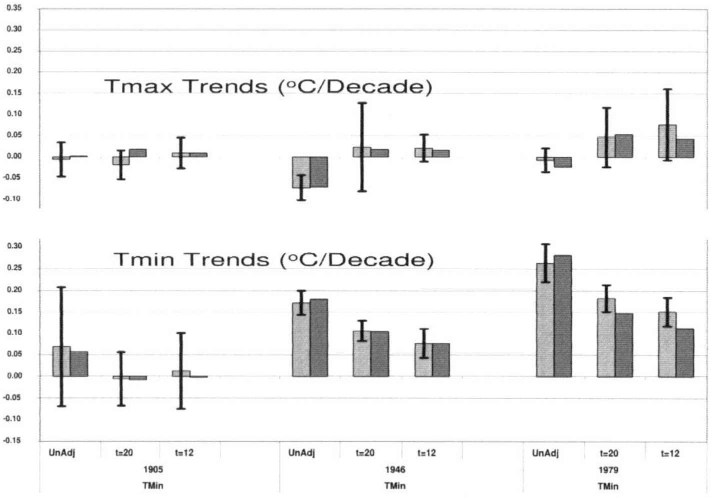

Figure 7. Trends in maximum and minimum temperature in the central cell of the East African data set [7].

3.1 Equation Set and Solution Techniques

Before outlining the proposed new investigations we outline the equation sets and techniques previously developed. We consider a simple, one-dimensional (single column model), see [27], [28], [29].

$$\frac{\partial u}{\partial t} = f_{co}(v - v_G) + \frac{\partial}{\partial z}(K_m(z,u_z,v_z,\theta_z)\frac{\partial u}{\partial z}), \tag{1}$$

$$\frac{\partial v}{\partial t} = f_{co}(u_G - u) + \frac{\partial}{\partial z}(K_m(z,u_z,v_z,\theta_z)\frac{\partial v}{\partial z}), \tag{2}$$

$$\frac{\partial \theta}{\partial t} = R_c(\theta) + \frac{\partial}{\partial z}(K_h(z,u_z,v_z,\theta_z)\frac{\partial \theta}{\partial z}), \tag{3}$$

here u,v are the wind components and Θ the potential temperature with the boundary conditions (for $t \geq 0$)

$$u(z_0,t) = 0, \qquad u(Z,t) = u_G,$$

$$v(z_0,t) = 0, \qquad v(Z,t) = v_G,$$

$$\theta(z_0,t) = \theta_g(t), \qquad \theta(Z,t) = \theta_Z,$$

coupled with the IVP for the ground potential temperature $\theta_g = \theta_g(t)$ (for z=0, t>0)

$$\frac{d\theta_g(t)}{dt} = \frac{1}{C_b}(I_1(\theta)|_{z=z_0} - \sigma\theta_g^4(t) - H_0(z,u_z,v_z,\theta_z)|_{z=z_0})$$
$$-\kappa_m(\theta_g(t) - \theta_m), \tag{4}$$

and the initial conditions

$$u(z,0) = u_G, \quad v(z,0) = v_G, \quad \theta(z,0) = \theta_A,$$

$$\theta_g(0) = \theta_A.$$

Here,

$K_m = K_m(z,u_z,v_z,\theta_z)$ and $K_h = K_h(z,u_z,v_z,\theta_z)$

are the exchange coefficients for momentum and heat, which are functions of the space variable z and the partial derivatives:

$$u_z = \partial u(z,t)/\partial z, \qquad v_z = \partial v(z,t)/\partial z, \qquad \theta_z = \partial \theta(z,t)/\partial z.$$

The nonlinearity in the equations enters through these terms, originally defined as

$$K_m = f_m(Ri)\kappa^2 z^2 s, \qquad K_h = f_h(Ri)\kappa^2 z^2 s,$$

with the stability functions f_m and f_h (namely, the Richardson number formulations) given by [26].

$$f_m(Ri) = f_h(Ri) = \begin{cases} (1 - Ri/Ri_c)^2, & if\, 0 \leq Ri \leq Ri_c, \\ 0, & if\, Ri \geq Ri_c, \end{cases}$$

and the vertical wind shear s defined by

$$s \equiv s(u_z, v_z) = \sqrt{u_z^2 + v_z^2},$$

$R_c(\theta)$ is the clear-air radiative cooling rate for the air temperature and was initially taken to be zero. I_1 is the longwave back radiation from the atmosphere, given by [30]

$$I_1 \equiv I_1(\theta)|_{z=z_0} = 0.67\sigma(1670Q_a)^{0.08}\theta^4|_{z=z_0} + GHGS,$$

and H_0 is the heat flux carried away from the surface by turbulence.

3.2 Truncated Two-Layer Model

The PDE set above is very close or identical to boundary layer formulations used in weather forecast and new generation global climate models [31]. The surface equation does not explicitly include vegetative effects. However, the bulk heat capacity can be considered that of a composite surface so that the bulk resistance to the combination of the soil and vegetation can be included [32]. In order to simplify the analysis of the time dependent and steady state behavior of the solutions, but still hoping to capture the essential physical and mathematical properties of the PDE system (1)-(4), we developed a simple 2 layer model by employing finite difference representation and Monin-Obukhov similarity theory. The space interval $z \in [z_0, Z]$ is divided into 2 layers: $z \in [z_o, z_1]$ and $z \in [z_1, z_2]$, with $z_2 = Z$. Denote the wind speed by $u_i = u(z_i), v_i = v(z_i)$ and the potential temperature by $\theta_i = \theta(z_i)$ at the first and second levels with $z = z_i, i = 1,2$, and the ground potential temperature $\theta_g =$ constant at the roughness level with $z = z_0$. With $u_2 = u_g, v_2 = v_g, \theta_2 = \theta_Z$ given and fixed, we are to solve for $u = u_1, v = v_1$ and $\theta = \theta_1$ at the first level $z = z_1$, and $\theta = \theta_g$ at the roughness level $z = z_0$. Throughout the "ground" layer $z \in [0, z_0]$, the potential temperature is assumed to be independent of the height z, and is a function of the time t only: $\theta_g = \theta_g(t)$; The wind field is identically zero: $u \equiv v \equiv 0$.

Applying the centered finite difference with respect to z , we reach a system of 4 first order ODE's for the two layer model:

$$\frac{du_1}{dt} = f_{co}(v_1 - v_G) + \frac{1}{z_{3/2} - z_{1/2}}\left(\frac{K_{m,3/2}(u_2 - u_1)}{z_2 - z_1} - u_*^2\cos(\psi)\right), \tag{5}$$

$$\frac{dv_1}{dt} = f_{co}(u_G - u_1) + \frac{1}{z_{3/2} - z_{1/2}}\left(\frac{K_{m,3/2}(v_2 - v_1)}{z_2 - z_1} - u_*^2\sin(\psi)\right), \tag{6}$$

$$\frac{d\theta_1}{dt} = \quad + \frac{1}{z_{3/2} - z_{1/2}}\left(\frac{K_{h,3/2}(\theta_2 - \theta_1)}{z_2 - z_1} - u_*\theta_*\right), \tag{7}$$

$$\frac{d\theta_g}{dt} = \frac{1}{C_b}\left(I_1 - \sigma\theta_g^4 - H_0\right) - \kappa_m\left(\theta_g - \theta_m\right), \tag{8}$$

with initial conditions $u_1(0) = u_G$, $v_1(0) = v_G$, $\theta_1(0) = \theta_A$. Where u_*, θ_* are the friction velocity and friction temperature. The above system can be utilized to examine the role parameters may have in changing the solutions in the stable boundary layer and minimum temperatures.

3.3 Role of Surface Parameters

As human population has expanded more land has been altered by urban, suburban and rural development. This development has altered parameters that can impact minimum temperatures and include roughness, heat capacity and conductivity of the surface. Before showing the dependence of GHG forcing on the air temperature in the SNBL, we will explain a bifurcation diagram with imposed geostrophic wind as the bifurcation parameter (Figure 8). In this case as geostrophic wind speed increases, air temperature transitions from a cold to a warm solution. The transition occurs rapidly near 7 m s^{-1}. This illustrates the competition between developing thermal stability and shear. In the cold solution surface cooling leads to thermal stability which inhibits turbulence and causes air temperature to become decoupled from the warm atmosphere aloft. As air temperature decreases, turbulence is further inhibited. Above a critical wind speed vertical wind shear dominates thermal stability and turbulence is generated causing warm air aloft to be mixed to the surface. At larger values of surface roughness the bifurcation diagram has a fold-back region, which indicates that multiple solutions are supported for the same parameter and thus sensitive to initial conditions. The existence of two states in the model is consistent with the radiative and turbulent classifications in observations [14, 17, 33]. In the present model the magnitude of the warming as the model transitions from the cold to warm solution is dependent on the layer depths chosen. However, here, as in [26] and other uses of this model, the layer depths are chosen so that the warming is consistent with that of actual surface temperatures in observations for turbulent and radiative nights [18, 33]. It is also consistent with the strength of the near-surface inversion [34] which would largely be destroyed in the turbulent case and consistent with the surface transition in the lowest 10m described by [17].

Surface roughness can also change the competition and the solution. Figure 9 shows the bifurcation diagram with roughness as the bifurcation diagram. It shows that at intermediate wind speeds there can be sharp increases in air temperature as the roughness increases. At most all wind speeds roughness increases near surface air temperature.

Surface heat capacity can also alter the behaviour of the surface. In the present model the bulk surface heat capacity is given capacity per unit area, C_b, is given by

$$C_b = \rho_g c_g \Delta d ,$$

where ρ_g is the soil density, c_g is the soil specific heat capacity, and Δd is the soil depth. Figure 10 shows the bifurcation diagram with heat capacity as the bifurcation parameter. For the light wind case (u_G = 8 ms^{-1}) the air and slab temperature show a strong dependence on bulk heat capacity, especially at values between 3000 and 8000 $Jm^{-2}K^{-1}$, where the air and surface temperature decrease as the bulk heat capacity decreases. In the highly stable regime turbulence is suppressed and the surface is dominated by long-wave loss. Thus, the smaller the bulk heat capacity the larger the loss. For larger geostrophic wind speeds (u_G = 16 ms^{-1}) the relationship changes. Now surface and air temperatures decrease with increasing heat capacity. In this regime turbulent heat transfer is strong and warms the surface. The larger the heat capacity becomes a resistance to change, thus, temperatures decrease for larger heat capacities. The intermediate wind-speed case, which shows multiple solutions for a range of heat capacities, is the most interesting. The two stable attracting solutions and one unstable solution are highly sensitive to initial conditions. We might even expect possible transition between states if perturbations were large enough.

3.4 Role of Radiative Forcing

GHG forcing has been postulated as a reason for nocturnal warming and decreasing DTR. Walters et al. [20] explored the sensitivity of the SBL to radiative forcing here we show only the intermediate wind case (figure 11). The bifurcation diagram is also shown for a range of forcing -300 to +150 W m^{-2} to demonstrate the shape of the bifurcation diagrams. However, these ranges are not to be construed as a relevant range of anthropogenic forcing. The S shape associated with larger roughness indicates that in some regions multiple solutions exist. Increased GHGS itself causes the system to transition from the cold to the warm solution. For rougher surfaces the diagram is S-shaped and supports multiple solutions. In practical terms, even slight changes in GHGS can lead to large changes in surface temperature (7-9 ^{o}C).

For simplifications of physical system such as for the two layer system above, there is concern whether the behaviour exhibited in the simple model will hold in more complete systems where additional degrees of freedom are available. We have begun to explore these issues using a time dependent column model (equations (1)-(4) above) and they will be fully reported later. In preliminary analyses of the time dependent model we found that we did not get as sharp a transition at the surface as found in the bifurcation diagrams of [20]. Rather, the sharp transition was found to occur near the top of the boundary layer. Of most importance we did find that the destabilization of the NBL by the enhanced long wave forcing was retained. And, in fact a substantial part of the warming at the surface was due to enhanced vertical mixing of warm air from aloft. These experiments are to be reported in 34. Figure 12 shows the difference in profiles after the control case was subjected to an additional downward radiation of 4 Wm^{-2}.

4. Issues with Global Scale Models and the Stable Boundary Layer

As indicated at the beginning, global scale models have in general not captured the magnitude of the differential trend in maximum and minimum temperatures. Part of the issue may be with the type closures used in coarse grid models that reduce the sensitivity and response to land use changes and/or radiative forcing exhibited in the previous section. This analysis may also suggest why GCMs have not captured the magnitude of the observed decreasing DTR. The processes described in the equations analyzed here are not well represented in the SNBLs of GCMs because Richardson numbers, though dimensionless, are a direct function of Δz. The coarse resolution in GCMs does not allow the direct application of the principles of turbulent kinetic energy closures. Steeneveld et al. [33] showed that fine resolution is required to capture the behavior of the SNBL. When a large-scale model such as a GCM or even a mesoscale model attempts to run with first-principle closures, modelers must cope with its tendency to crash to a very cold solution [31, 35]. Thus, modelers generally add processes to produce mixing, such as altering profile functions [31], increasing roughness or making the critical Richardson number dependent on grid spacing [27]. These fixes may damp the type of sensitive transition shown here represented in the equations.

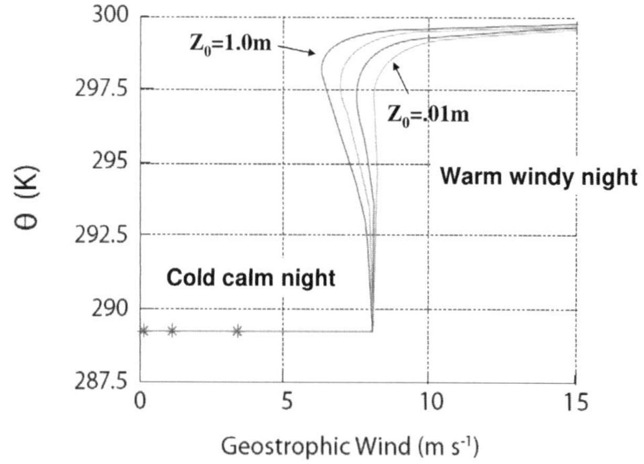

Figure 8. Bifurcation diagram for first layer temperature. Bifurcation parameter is geostrophic wind.

IOP Conf. Series: Earth and Environmental Science **13** (2010) 012003 doi:10.1088/1755-1315/13/1/012003

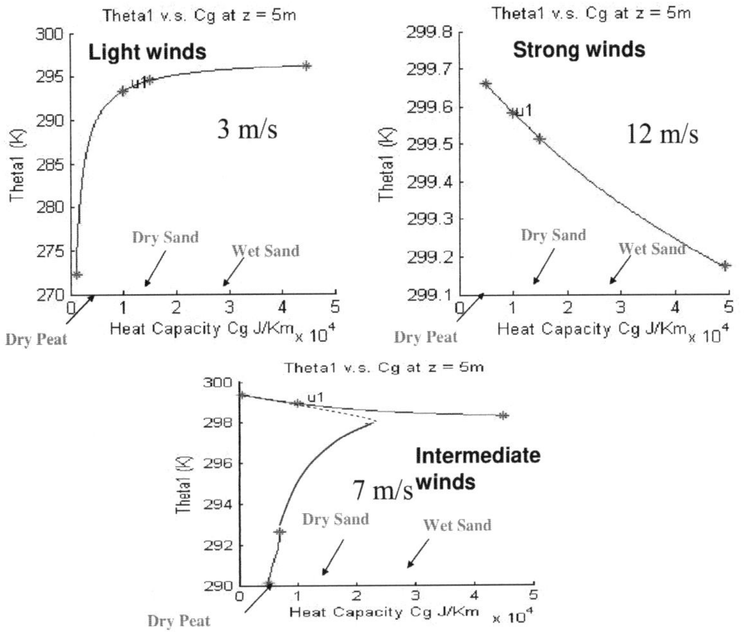

Figure 9. Bifurcation diagram for first layer temperature. Bifurcation parameter is bulk heat capacity (C_B).

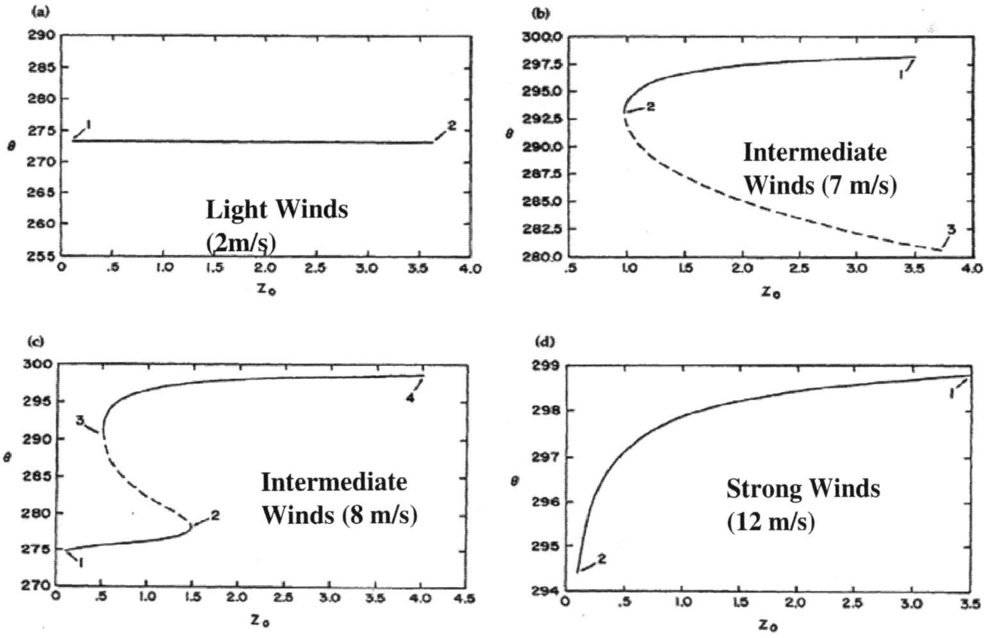

Figure 10. Bifurcation diagram for first layer temperature. Bifurcation parameter is surface roughness (z_o).

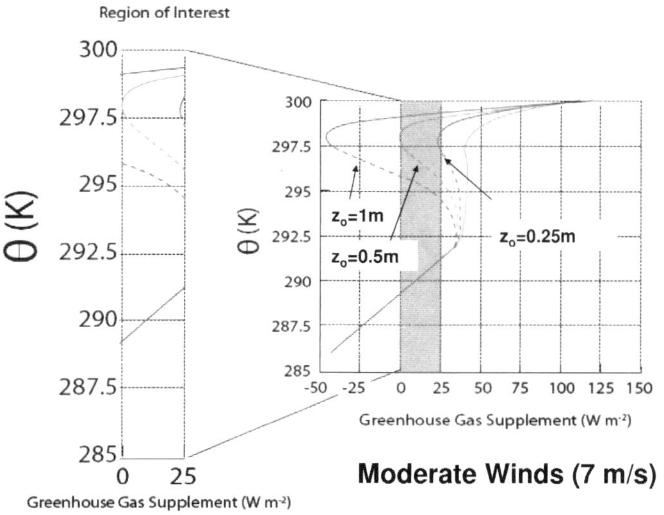

Figure 11. Bifurcation diagram for first layer temperature. Bifurcation parameter is greenhouse gas supplement [20].

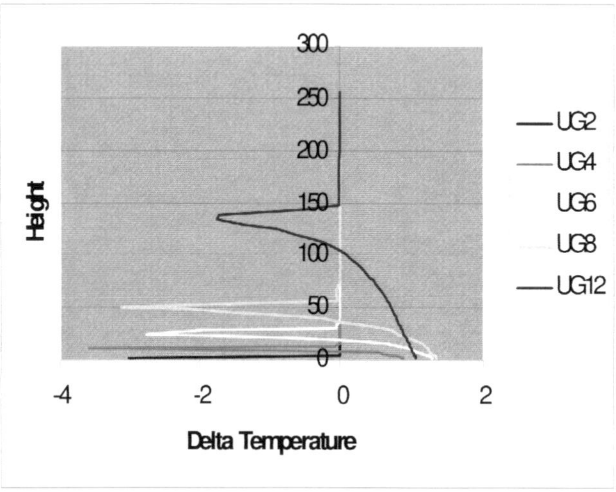

Figure 12. Vertical profile of warming difference due to added 4 Wm^{-2}. Note a significant part of the warming is due to redistribution of heat from aloft.

5. The present state of main cryosphere components

Minimum temperatures in the stable boundary layer have increased more rapidly than maximum temperatures in the historical record. The fact that the difference in trends in T_{Max} and T_{Min} continues and in fact accelerates in the period 1979-2004 in East Africa may be important in interpreting the results of [6]. While it is possible that East Africa difference trends are indeed different than that of the globe as provided by [6], there is concern that the reduced number of stations in the 1979-2004 GHCN data set may not be sampling many of the areas of the globe that are behaving like East Africa. Thus, it is important that data sets such as GCHN that separate minimum and maximum temperatures be expanded to include more stations distributed around the globe.

The night-time boundary layer is very shallow and by its very existence is decoupled from the deeper atmosphere. Thus, the stable nocturnal boundary layer does not represent the heat content in a large part of the atmosphere. Because of non-linearities in the stable boundary layer in some parameter spaces, T_{min} is highly sensitive to slight changes in cloud cover, greenhouse gases, and other radiative forcing. This is consistent with the observation studies such as [36] and theoretical studies of [37]. However, the analyses above indicate this sensitivity is reflective of a change in the turbulent state of the atmosphere and a redistribution of heat not a change in the heat content of the atmosphere. Because of the redistribution phenomena and the shallow layer affected, observed minimum temperatures are a very poor measure of the accumulation of heat in the deep atmosphere that is of major concern in greenhouse gas induced climate change.

Climate models with their course resolution cannot accurately simulate the physics important to heat redistribution, their minimum temperatures are suspect. Surface maximum temperatures would seem to represent a more robust measure of the heat content of the atmosphere since daytime boundary layers connect the surface to a depth of one to two kilometres or more and are well mixed. Climate models more accurately simulate daytime mixing which, by its non-local nature, is not as dependent on grid resolution. Thus, daytime maximum temperatures should be much better for detecting and simulating this heat accumulation in the deep atmosphere. However, minimum temperatures may be accurately reflecting local climate change and may be important to societal impacts and to interactions with the cryosphere.

References

[1] Karl, T. R., P. D. Jones, R. W. Knight, G. Kukla, N. Plummer, V. Razvayev, K. P. Gallo, J. Lindseay, J. R. J. Charlson, T. C. Peterson, 1993. Asymmetric trends in surface temperature, *Bull. Am. Meteorol. Soc.*, **74**, 1007-1023.

[2] Hansen, J., R. Ruedy, J. Glascoe, and M. Sato, 1999. GISS analysis of surface temperature change. *J. Geophys. Res.*, **104**, 30997-31022, doi:10.1029/1999JD900835.

[3] Brohan, P., J. J. Kennedy, I. Harris, S. F. B. Tett, and P. D. Jones, 2006. Uncertainty estimates in regional and global observed temperatures changes: A new data set from 1850. *J. Geophys. Res.*, **111**, D12106, doi:10.1029/2005JD006548.

[4] Peterson, T. C. and R. S. Vose, 1997. An overview of the Global Historical Climatology Network temperature database. *Bull. Amer. Meteor. Soc.*, **78**, 2837-2849.

[5] Easterling, D. R., B. Horton, P. D. Jones, T. C. Peterson, T. R. Karl, D. E. Parker, M. J. Salinger, V. Razuvayev, N. Plummer, P. Jamason, and C. K. Folland, 1997. Maximum and minimum temperature trends for the globe. *Science*, **277**, 364-367.

[6] Vose, R. S., D. R. Easterling, and B. Gleason, 2005. Maximum and minimum temperature trends for the globe: An update through 2004, *Geophys. Res. Lett.*, **32**, *L23822.*

[7] Christy, J.R., William B. Norris, and Richard T. McNider, 2009. Surface Temperature Variations in East Africa and Possible Causes, *J. Climate*, **22**, 3342–3356 DOI: 10.1175/2008JCLI2726.

[8] Cao, H. X., J. F. B. Mitchell, and R. Lavery, 1992. Simulated diurnal range and variability of surface temperature in a global climate model for present and doubled CO_2 climates, *J. Climate*, **5**, 920-943.

[9] Stone, D. A. and A. J. Weaver, 2003. Factors contributing to diurnal temperature range trends in the twentieth and twenty first century simulations of the CCAma coupled model, *Climate Dynamics*, **20**, 435-445.

[10] Dai, A., K. E. Trenberth, and T. R. Karl, 1999. Effects of clouds, soil moisture, precipitation and water vapor on diurnal temperature range, *J. Climate*, **12**, 2451-2473.

[11] Durre, I. and J. Wallace, 2001. Factors influencing the cold season diurnal temperature range in the United States, *J. Climate*, **14**, 3263-3278.

[12] Travis, D.J., A. Carleton, R. Lauritsen, 2004. Regional variations in U.S. diurnal temperature range for the 11-14 September 2001 aircraft groundings: Evidence of jet contrail influence on climate. *J. Climate*, **17**, 1123 -1134.

[13] Christy, J. R., W. B. Norris, K. Redmond, and K. P. Gallo, 2005: Methodology and results of calculating Central California surface temperature trends: Evidence of human-induced climate change?, *J. Clim.*, **19**, 548– 563, doi:10.1175/JCLI3627.1.

[14] Van de Wiel, B. J. H., R. J. Ronda, A. F. Moene, H. A. R. De Bruin, and A. A. M. Holtslag, 2002a. Intermittent turbulence and oscillations in the stable boundary layer over land. Part I: A bulk model, *J. Atmos. Sci.*, **59**, 942-958.

[15] Van de Wiel, B. J. H., A. F. Moene, R.J. Ronda H. A. R. De Bruin, and A. A. M. Holtslag, 2002b. Intermittent turbulence and oscillations in the stable boundary layer over land. Part II. A system dynamics approach, *J. Atmos. Sci.*, **59**, 2567-2581.

[16] ReVelle D. O., 1993. Chaos and "bursting" in the planetary boundary layer, *J. Appl. Meteor.*, **32**, 1169–1180.

[17] Mahrt, L., 1999. Stratfied atmospheric boundary layers, *Bound-layer Meteor.*, **90**, 375-396.

[18] Acevedo, O. C. and D. R. Fitzjarrarld, 2001. The early evening surface layer transition: Temporal and spatial variability. *J. Atmos. Sci.*, **58**, 2650-2667.

[19] Doedel, E.J., H.B. Keller, and J.P. Kernevéz, 1991. Numerical analysis and control of bifurcation problems: Bifurcation in finite dimensions, *Int. J. Bif. & Chaos*, **1**, 493-520.

[20] Walters, J. T., R. T. McNider, X. Shi, and W. B. Norris, 2007. Positive surface temperature feedback in the stable nocturnal boundary layer. *Geophys. Res. Let.*, doi:10.1029/2007/GL029505.
http://vortex.nsstc.uah.edu/public/outgoing/biazar/nsf2009/Walters_et_al_GRLCopy.pdf

[21] **NRC**, 2000. *Reconciling Observations of Global Temperature Change*. National Research Council, Board on Atmospheric Sciences and Climate, Commission on Geosciences, Environment and Resources, National Academy Press, Washington D.C., 85 pp.

[22] Karl, T. R., S. J. Hassol, C. D. Miller, W. L. Murray (eds), 2006. Temperature trends in the lower atmosphere: Steps for understanding and reconciling differences. A report by the Climate Change Science Program (CCSP) and the Subcommittee on Global Change Research, Washington, DC, US Government. (Report at http://www.climatescience.gov/Library/sap/sap1-1/finalreport/default.htm)

[23] Klotzbach, P. J., R. A. Pielke, R. A. Pielke, J. R. Christy, and R. T. McNider, 2009. An alternative explanation for differential temperature trends at the surface and in the lower troposphere, *J. Geophys. Res.*, **114**, D21, doi:10.1029/2009JD011841

[24] Pielke, R.A., Sr. and T. Matsui, 2005. Should light wind and windy nights have the same temperature trends at individual levels even if the boundary layer heat content is the same? *Geophys. Res. Lett.,* **32**, L21813:doi 10.1029/2005GL024407.

IOP Conf. Series: Earth and Environmental Science **13** (2010) 012003 doi:10.1088/1755-1315/13/1/012003

[25] Blackadar, A. K., 1979. High resolution models of the planetary boundary layer, *Adv. Environ. Sci. Eng.,* **1**, 50- 85.

[26] McNider, R. T., X. Shi, M. Friedman, and D. E. England, 1995. On the predictability of the stable atmospheric boundary layer, *J. Atmos. Sci.,* **52**, 1602-1614. http://vortex.nsstc.uah.edu/public/outgoing/biazar/nsf2009/McNider_Predictability.pdf

[27] McNider, R. T. and R. A. Pielke, 1981. Diurnal boundary layer development over sloping terrain, *J. Atmos. Sci.*, **38**, 2198-2212.

[28] Stull, R. B., 1988. An Introduction to Boundary Layer Meteorology. Kluwer Academic Publishers. ISBN: 978-90-277-2769-5680.

[29] Shi, X., McNider, R. T., D. E. England, M. J. Friedman, W. Lapenta, and W. B. Norris, 2005. On the behavior of the stable boundary layer and role of initial conditions, *Pure Appl. Geophys.*, **162**, 1811-1829.

[30] Staley, D.O. and G. M. Jurica, 1972. Effective Atmospheric Emissivity under Clear Skies, *J. Appl. Meteorol.*, **11**, 349-356.

[31] Viterbo, P., A. Beljaars, J. F. Mauhoff, and J. Teixeira, 1999. The representation of soil moisture freezing and its impact on the SBL, *Quart. J. Roy. Met. Soc.*, **125**, 2401-2426.

[32] McNider, R. T., W.M. Lapenta, A. Biazar, G. Jedlovec, R. Suggs, and J. Pleim, 2005. Retrieval of grid scale heat capacity using geostationary satellite products: Part I: Case-study application, *J. Appl. Meteor.,* **88**, 1346-1360.

[33] Steeneveld, G. J., B. J. H. van de Wiel, and A. A. M. Holtslag, 2006. Modeling the evolution of the atmospheric boundary layer coupled to the land surface for three contrasting nights in CASES-99, *J. Atmos. Sci.*, **63**, 920–935.

[34] McNider, R.T., A. Biazar, F. Shi, M. Friedman, J. Hughes, I. Savin, 2009. Positive temperature feedback in a multi-level high resolution model of the nocturnal boundary layer subjected to climatic radiative forcing. To be submitted to *J. Geophy. Res.*

[35] Derbyshire, S. H., 1999: Boundary-layer decoupling over cold surfaces as physical boundary instability, *Boundary-layer Meteorology*, **90**, 297-325.

[36] Runnalls, K. E. and T. R. Oke, 2006. A microclimatic technique to detect inhomogeneities in historical records of screen-level air temperature. *J. Climate*, **19**, 959-978.

[37] Esau, I., 2008: Formulation of the planetary boundary layer feedback in the Earth's climate system, *Computational Technologies,* **13**, 95-103.

IOP Conf. Series: Earth and Environmental Science **13** (2010) 012004 doi:10.1088/1755-1315/13/1/012004

Contrasting the surface ocean distribution of bromoform and methyl iodide; implications for boundary layer physics, chemistry and climate

C. J. Palmer [1]

Department of Oceanography, University of Cape Town, 7701, South Africa

carl.j.palmer@gmail.com

Abstract. Bromoform and methyl iodide are both methane-like hydrocarbons with a halogen atom replacing one or more of the hydrogen atoms. Both of these compounds occur naturally in the environment as a result of their production from seaweed and kelp. They are of interest to climate science as a result of their catalytic destruction of boundary layer ozone (a potent greenhouse gas) and, specifically for methyl iodide, the proposed role in the formation of new cloud condensation nuclei with implications for climate. In this paper, the currently available data on the distribution of bromoform and methyl iodide are analysed and contrasted to show that the concentrations of bromoform and methyl iodide do not correlate, that, in contrast to bromoform, the parameterization of sea surface methyl iodide concentrations demands only the sea surface temperature, and that the pelagic distribution of methyl iodide appears to follow the solar zenith angle. These three observations together suggest that, while the pelagic source of bromoform is mostly biogenic, the source of methyl iodide is photochemical. This has implications for the understanding of planetary boundary layer chemistry and potential organohalogen mediated feedbacks to climate.

1. Introduction

Iodine and bromine exist in the oceans mainly in the form of non-volatile and soluble dissolved ions at approximately 0.06 and 70 ppm respectively [1]. These halides are selectively incorporated into algal cells, a fact that has been known for over one hundred years [2], and converted to organic halides (organohalogens). Bromoform and methyl iodide are two naturally occurring organohalogens that are produced in the Marine Boundary Layer (MBL – defined here as encompassing the surface ocean and the part of the planetary boundary layer that lies above it) by phytoplankton [3,4], ice algae [5] and kelp [e.g. 6-10]. Importantly, unlike the inorganic iodide and bromine compounds, these organogalogens are volatile and sparely soluble and are therefore readily outgassed to the MBL [11].

Once in the MBL, these organohalogens photolyse to produce halogen radicals, with important implications for atmospheric scientists: initial interest was sparked by the observation that bromine radical species can lead to catalytic MBL ozone (O_3) depletion [12], thus perturbing the oxidative potential of the atmosphere. Furthermore, bromoform may persist long enough in the atmosphere to reach the stratosphere and affect the ozone layer [13,14]. Due to rapid convection in the tropics, recent modeling studies and measurements suggest that bromoform contributes significantly (30 to >60%) to inorganic bromine levels within the upper troposphere (UT)-lower stratosphere (LS) region [14, 15-17]. This represents an important uncertainty in climate science, as in this region ozone is a significant greenhouse gas [18]. Methyl iodide produces iodine radicals that, as well as also catalytically destroying ozone, have been implicated in the formation of cloud condensation nuclei, with consequences for climate [19]. Furthermore, organo-iodide compounds represent the only known mechanism by which iodine, an essential component of our diet, reaches the continents.

The reasons for the biological accumulation of bromine and iodine in algae cells remained a mystery for almost a centaury after its discovery, until recent work suggesting that these halides may have a use in algal cells as primitive antioxidants [20]. Iodine and bromine may react with oxidants such as hydrogen peroxide (H_2O_2) or ozone to produce halogen oxides, which in turn may react with dissolved organic matter via the haloform reaction to produce organohalogens. Thus, the production of organohalogens is thought to be as an indirect consequence of this metabolic response to oxidative stress [20, 21]. Production rates are greatest in kelps and around two orders of magnitude less in phytoplankton. However, with the distribution of the former being limited to coasts, the dominant source still remains unclear.

Table 1. Comparison of physical data and biological production rates for bromoform and methyl iodide

	Bromoform ($CHBr_3$)	**Methyl Iodide (CH_3I)**
Molecular weight	252.73 g mol^{-1}	141.94 g mol^{-1}
Solubility in water	3.2 g l^{-1} (at 30 °C)	0.14 g l^{-1} (at 20 °C)
Boiling point	149.1 °C	42.4 °C
Typical atmospheric lifetime [11,22]	Days – months	Days - months
Typical macro-algal production rates [11]	ca. 1000 pmol g^{-1} (fw) hr^{-1}	10-100 pmol g^{-1} (fw) hr^{-1}
Typical phytoplankton production rate [23,24]	ca. 10 pmol g^{-1} (fw) hr^{-1}	[a]ca. 10 pmol g^{-1} (fw) hr^{-1}

[a] Fresh weight was estimated from measurements per cell number by assuming the average cell size is 100 μm, the density of the cell is the same as silica (density 2.634 g cm^{-3}) and that the fresh weight is four times the dry weight.

Whereas there are some known anthropogenic sources of bromoform (e.g. water chlorination and rice production), these sources are far smaller than natural sources when considered on a global scale [24]. There is some evidence to suggest that there is a locally significant terrestrial source of bromoform to the troposphere off the North-West coast of Africa [25] but this source has neither been identified nor quantified. Methyl iodide, on the other hand, has been shown to be produced in laboratory studies by an abiotic photochemical mechanism [26]. However, this is generally considered as an interesting observation, but not a significant addition to the biological sources [27]. In this paper, we review the contrasting observational and model evidence to show that whereas the environmental distribution of bromoform is indicative of a biological source, the distribution of methyl iodide strongly suggests a dominant abiotic source over much of the pelagic oceans.

2. Methods

Field measurements from approximately 50 previous studies of bromoform were collated as described by Palmer and Reason [28] and this process was repeated for di-bromomethane and methyl iodide, with the latter data set incorporating additional studies where only methyl iodide was measured [29,30]. For each bromoform or methyl iodide data point, auxiliary data, including the estimated mixed layer depth, Sea-WiFS chlorophyll-*a*, sea surface temperature, salinity, and the QuickCOW monthly wind speed, was collated as described for bromoform by Palmer and Reason [28]. In addition to these data, shipboard measurements of air temperature, distance from nearest coast, wind speed and precipitation were also collated for each organohalogen point where available.

A climatology of methyl iodide surface ocean concentrations was developed by testing model parameterization formulas using a method based on that presented by Palmer and Reason [28]. The method presented here departs from the published method however, in that the model formulas to be tested were not manually chosen; instead, all possible linear combinations of parameterization variables were tested sequentially and the most successful model formulas were evaluated based on minimizing the root mean squared deviation (RMSD) between observed and modelled. Identifying the most successful model formula in this manner represents a non-deterministic polynomial time - complete (NP-C) style problem. However, the number of permutations (ca. 2 x 10^8) is limited enough that this problem can be solved with a typical desktop computer in 3-5 days.

3. Contrasting Field Observations

Previous studies provide comprehensive reviews of existing measurements of both bromoform [24] and methyl iodide [31] and therefore this study does not intend to reproduce these reviews, but rather to use these, along with the data set of Palmer and Reason [28] and more recent work, to draw conclusions from the contrast in the distributions of these two organohalogens.

The most abundant of all organohalogens in surface seawaters is thought to be bromoform [24, and references therein]. Maximum reported surface water concentrations in any given study range from 1 to 4500 pmol l^{-1}, the highest concentrations being reported in the Canary Islands [32]. Unlike the other predominantly biogenic organohalogens, there have been no studies showing bromoform concentrations to be under the detection limit, suggesting that this compound is more ubiquitous throughout the oceans than methyl iodide, which has been observed below limits of detection, even when this limit has been less than 1 ppt [33]. In both the Pacific [34] and Atlantic [35] Oceans, concentrations of bromoform have been shown to be correlated with chlorophyll-*a* concentrations. This is in contrast to methyl iodide which has been show to be correlated only with sea surface temperature [29,30].

Common ground can however be found when comparing the difference between coastal and pelagic concentrations of these two compounds as both exhibit significantly higher coastal concentrations. In temperate and subtropical regions, the surface seawater concentration of bromoform in coastal regions is over 100 times higher than in the open ocean [24]. For example, Class *et. al.* [36] reported mean open ocean concentrations at 33 °N, 14 °W to be 3.17 pmol l^{-1} whereas coastal concentration at a similar latitude (Canary Islands; 28 °N, 17 °W) averaged 2000 pmol l^{-1} [32]. Similarly, methyl iodide is also found in vastly higher concentrations in coastal regions. For example, concentrations of around 15 ppt were recorded off the coast of Ireland [6] and Nova Scotia [37] compared to less than 1 ppt in the corresponding open oceans [31].

Several studies have reported high open ocean concentrations of bromoform in tropical regions, with a peak of ca. 2 pmol l^{-1} [28] at ~10 °N in the Atlantic and up to 3 pmol l^{-1} [37] in the equatorial Pacific. There were no significant differences between reported concentrations of bromoform in the open oceans of the Atlantic and Pacific [28,37]. Methyl iodide in contrast appears to remain at less than 1 ppt throughout most of the open ocean [31], with only one study reporting much higher concentrations [29]. Importantly when open ocean concentrations of bromoform and methyl iodide are compared, there is no significant correlation ($R^2 \approx 0$), whereas when bromoform is compared to another compound of predominantly biogenic origin (di-bromomethane) a significant correlation exists ($R^2 = 0.72$; Figure 1).

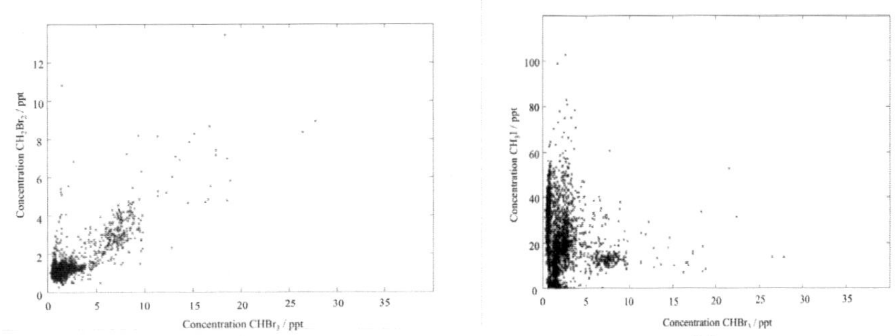

(a) **(b)**

Figure 1. Graph to show the relationship between surface ocean concentrations of **(a)** bromoform ($CHBr_3$) and di-bromomethane (CH_2Br_2), n = 2100, $R^2 = 0.72$ and **(b)** bromoform and methyl iodide (CH_3I), n= 2900 $R^2 = -0.00041$.

4. Climatology Studies

It has previously been demonstrated that the surface ocean bromoform concentrations in tropical water show a correlation with salinity, chlorophyll-*a* and mixed layer depth [28]. Testing all linear combinations of parameterization variables to model the sea surface concentrations of bromoform on a global scale yielded results that showed no significant improvement on the climatology of Palmer and Reason [28], with multiple solutions yielding an R^2 and RMSD of similar magnitude to that previously published. However, this paper is not intended as an in depth discussion of recent improvements in the parameterization of sea surface bromoform concentrations and the specifics of these solutions are therefore not discussed; it is sufficient to note that the implementation of the method described here to bromoform yields no significant improvement on previous work.

The same method of testing all possible linear combinations of parameterization variables applied to methyl iodide surface ocean concentrations yields very different results, however. The only parameterization variable required to minimize the RMSD was the sea surface temperature, with the model equation,

$$[CH_3I] = 8.8 + 0.93 * SST, \qquad (1)$$

where $[CH_3I]$ is the methyl iodide surface ocean concentration (ppt) and *SST* denotes the sea surface temperature (°C). A comparison of modeled concentrations (based on equation 1) against observed concentrations resulted in a RMSD of 11.4 and an R^2 of 0.25 (Figure 2).

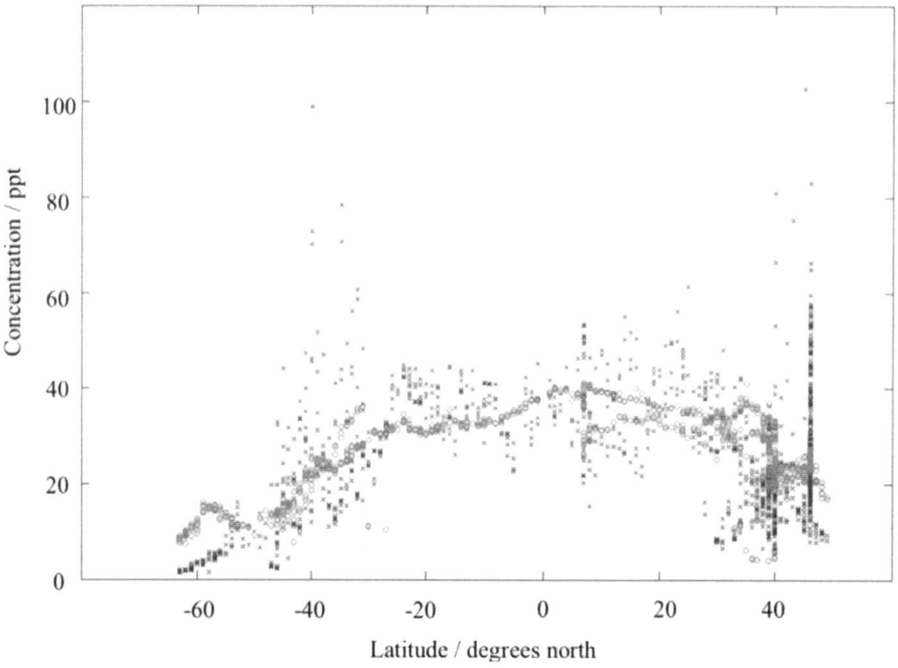

Figure 2. Relationship between methyl iodide and latitude for collated observations (denoted by blue crosses) and modelled based on the parameterisation of sea surface temperature (denoted by orange circles). The R^2 correlation coefficient for the observed against the modelled data is 0.3 and the RMSD is 11.5.

5. Discussion

The observations presented in section three (Contrasting Field Observations) show that there is clearly some common ground in the patterns displayed in the distribution of bromoform and methyl iodide; specifically, they both increase an order of magnitude or more towards the coast. This is to be expected in light of the information in Table 1; coastal kelp is clearly the most significant source of both organohalogens. Furthermore, they both have similar atmospheric lifetimes. However, these data demonstrate clearly that the pelagic distributions of bromoform and methyl iodide are not similar and correlate very poorly (Figure 1). However, bromoform, a compound that has been shown to be produced biologically, correlates very well with another predominantly biogenic compound, di-bromomethane. Given that these compounds have similar properties and similar atmospheric lifetime, these data appear to imply that the pelagic sources of bromoform and methyl iodide are not the same. From these observations alone it can be inferred that, either the biological mechanisms by which bromoform and methyl iodide are produced are different, or that one of the two has a predominantly non-biological source. It should be noted that it is indeed likely that the two biological methods are different, with haloperoxidase enzymes being shown to be involved in bromoform and di-bromomethane production but not in the production of methyl iodide [38]. Thus, this result may be explicable while retaining the hypothesis of a predominantly biological source for both compounds.

The data presented in section four (Climatology Studies) show that, despite all permeations of regression formula based on a wide range of biologically relevant variables being tested, the method failed to produce a highly skilled climatology of methyl iodide, whereas the same method results in skillful prediction of bromoform concentrations. This is clearly illustrated by comparison of the two model equations correlation with the observed data; analysis of bromoform observations against the model of Palmer and Reason [28] yielded an R^2 correlation coefficient of 0.41 compared to that presented here for methyl iodide of 0.25. Furthermore, whereas several biologically relevant variables (e.g. chlorophyll-*a* concentration and mixed layer depth) were required to make an accurate prediction of bromoform surface ocean concentrations, only a physical variable (sea surface temperature) was required for the methyl iodide model formula. Whereas it has previously been shown that methyl iodide correlates with sea surface temperatures in individual studies [33, 34], Equation (1) represents the first attempt to formalize this relationship on a global scale. While this relationship appears to predict current methyl iodide concentrations with some skill, it is important to reflect that correlation does not imply causality. Also pertinent to note is that methyl iodide is positively correlated with temperature; a negative correlation may be explicable in terms of biological source due to the correlation of primary productivity and cold upwelled waters. These observations therefore strongly suggest that, whereas the predominant source of bromoform is biogenic, the source of methyl iodide is abiotic.

Further evidence for a predominantly abiotic source of methyl iodide is observed if the methyl iodide data is presented as a 2° running mean by latitude (Figure 3). These data show that the distribution of methyl iodide concentration displays two distinct peaks at around 30° North and 30° South respectively. The pattern of methyl iodide distribution reveals that its concentration correlates well with sea surface temperatures, as would be expected from the results of the regression analysis. However, if the data were then further disaggregated by season (Figure 4), the strongest peak is observed at 30° S in the boreal winter (DJF), 30° N in the boreal summer (JJA) and plateaus between these extremes in boreal autumn (the spring data is too sparse to make any coherent observations).

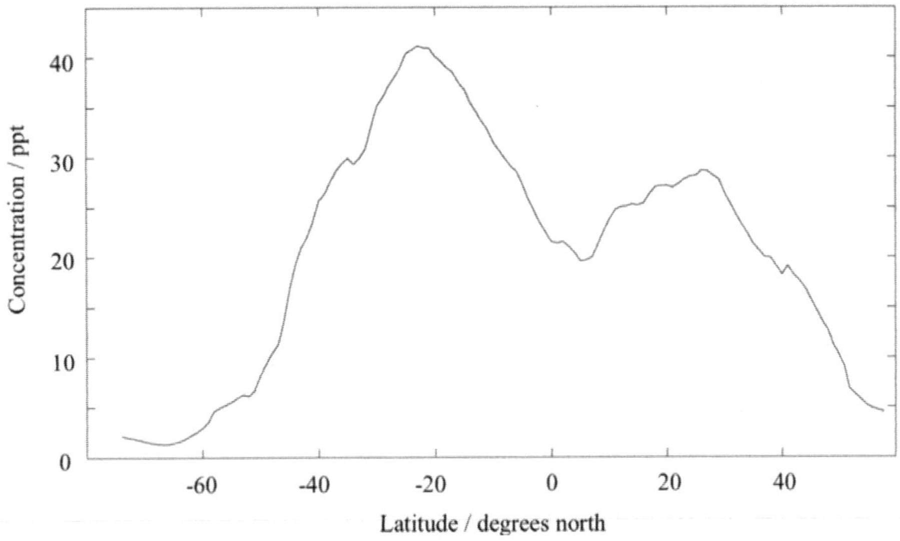

Figure 3. Two degree running average of methyl iodide concentrations by latitude with North and South edges treated by padding with the first and last value respectively.

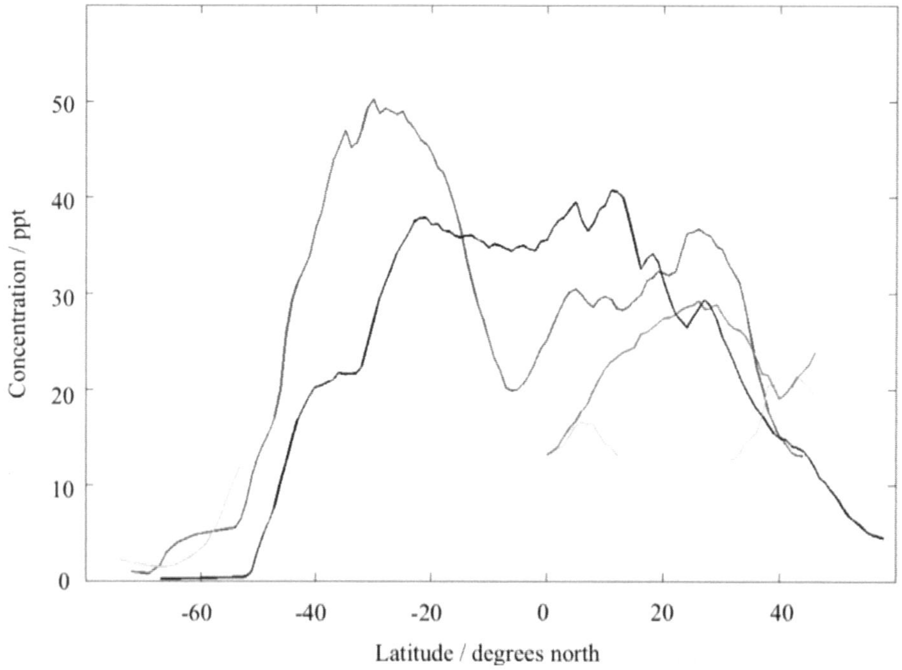

Figure 4. Two degree running average of methyl iodide concentrations by latitude and by season. Blue line denotes boreal winter (DJF), green line denotes spring (MAM), red line denotes summer (JJA) and black line denotes autumn (SON).

This change of peak methyl iodide distribution with season appears to follow the maximum solar zenith angle. These data therefore, appear to suggest that that predominant open ocean source of methyl iodide is in fact photochemical not biological as previously assumed. Considered together the data presented here suggest the following hypothesis: the general trend observed in pelagic surface ocean methyl iodide concentrations with latitude is a result of photochemical production, whereas points lying significantly above this are a result of enhanced production by biology. This hypothesis explains both the observed link with sea surface temperatures and the reason for our inability to fit model equation to the surface concentrations using biologically relevant variables.

It is important to note that this hypothesis has key implications for the chemistry of the MBL. As is discussed in the introduction, the methyl iodide that reaches the MBL from the oceans has an impact on the oxidative photochemistry and potentially on climate. The relationship between methyl iodide concentrations and sea surface temperature has led previous studies to propose a possible feedback mechanism where warmer oceans lead to more methyl iodide production, which in turn affects climate. If this is true then the relationship presented here may provide useful predictions to feed into these studies. However, if the hypothesis outlined above proves correct, then the correlation between sea surface temperature and methyl iodide does not prove causality; the real cause is solar radiation of which warmer surface oceans are also a byproduct. The real effect of a warmer climate on methyl iodide budgets would all depend on the effect global warming has on cloud cover; an issue of great uncertainty. Other studies have sought to demonstrate a link between CO_2 concentrations, ocean acidification and the biological production of organohalogens [39] in order to establish a feedback to climate. The results here suggest that making inferences about effects of an increase in CO_2 on methyl iodide production from algae should be treated with caution.

More work is however needed to demonstrate that the methyl iodide producing reactions are indeed photochemical and that light is indeed the limiting factor. For example, the observation also neatly follows the observed distribution of inorganic iodine species in seawater [40] and could also reflect a temperature dependent mechanism; modelling experiments may be able to provide data to better separate these possible mechanisms and limitations [41].

6. Conclusions

The data presented here demonstrate that, in coastal regions where kelp are abundant, it is likely that this represents the dominant source of both the organohalogens (methyl iodide and bromoform) in this case study. However, away from the coast, this appears not to be the case, with a biological source dominating for bromoform but an abiotic source dominating for methyl iodide. These results clearly demonstrate a fundamental difference in the distribution of the sea surface concentrations of bromoform and methyl iodide. The key observations central to this argument can be summarised as follows: (i) methyl iodide surface ocean concentrations show no correlation with those of bromoform, whereas bromoform does correlate with other biogenic chemicals, (ii) whereas bromoform is successfully modeled using a relationship that takes into account biologically pertinent variables, such as chlorophyll-*a* concentrations, methyl iodide is modeled using just an abiotic variable, (iii) our skill in parameterizing methyl iodide concentrations is significantly poorer than for bromoform, and (iv) analysis of methyl iodide concentration appears to reveal a link with the solar zenith angle. This is in contrast to bromoform, which displays a distribution that appears to be biologically driven.

The following hypothesis is thus proposed: unlike bromoform, the predominant pelagic source of methyl iodide is abiotic and most likely to be photochemical. This hypothesis is in contrast to previous studies, which assumed that, although a photochemical mechanism has been proved, it is only a minor contributor when compared to the biological source. This hypothesis thus raises important considerations for those studies looking to demonstrate a biogeochemical feedback to climate mediated by methyl iodide's biological production or its observed link to sea surface temperature. However, further work is necessary to differentiate a photochemical source from other abiotic sources, or biological sources with a physical control.

Acknowledgements
The National Research Foundation (NRF) of South Africa is gratefully acknowledged for funding this work. Many thanks for the useful comments of Brett Kuyper and Janine Basson, and to the all the participants at the International Conference on the Planetary Boundary Layer for their feedback.

References

[1] Segar D A 2007 *Introduction to Ocean Sciences* (W Norton and Company, Inc)
[2] Courtois B 1813 *Ann. Chim. (Paris)* **88** 304
[3] Tokarczyk R and Moore R M 1994 *Geophysical Research Letters* **21** 285-288.
[4] Carpenter L J, Wevill D J , Palmer C J and Michels J 2006 *Mar. Chem.* **3-4** 227
[5] Sturges W T, Cota G F and Buckly P T 1192 *Nature* **358** 660-662
[6] Carpenter L J, Malin G, Liss P S and Küpper F C 2000 *Global Biogeochemical Cycles* **14**, 1191
[7] Pedersen M, Collen J, Abrahamsson K and Ekdahl A 1996 *Scientia Marina* **60** 257
[8] Nightingale P D, Malin G and Liss P S 1995 *Limanology and Oceanography* **40** 680
[9] Manley S L, Goodwin K and North W J 1992 *Limnology and Oceanography* **37** 1652
[10] Gschwend P M and Macfarlane J K 1985 *Proceedings of the American Chemical Society* 189 **16**
[11] Carpenter L J and Liss P S 2000 *Journal of Geophysical Research* **105** 20539
[12] Barrie L A, Bottenheim J W, Schnell R C, Crutzen P J and Rasmussen R A 1988 *Nature* **334** 138
[13] Levine J G, Braesicke P, Harris N R P, Savage N H and Pyle J A 2007 *J. Geophys. Res.* **112** 15
[14] Sturges, W T, Oram D E, Carpenter L J and Penkett S A 2000 *Geophys. Res. Lett.* **27**, 2081
[15] Dvortsov, V L, Geller M A, Soloman S, Schauffler S M, Atlas E L and Blake D R 1999 *Geophys. Res. Lett.* **26** 1699
[16] Nielsen J E and Douglass A R 2001 *J. Geophys. Res.* **106** 8089
[17] Shauffler S M, Atlas E, Flocke F, Lueb R A, Stroud V and Travnicek W 1998 *Geophys. Res. Lett.* **25** 317
[18] Yang X, Cox R A, Warwick N J, Pyle J A, Carver G D, O'Connor F M and Savage H N 2005 *J. Geophys. Res.* **110**, 1029
[19] O'Dowd C D, Geever M, Hill M K, Smith M H, Jennings S G 1998 *Geophysical Research Letters*, **25**, 1661
[20] Küpper F C, Carpenter L J, McFiggans G B, Palmer C J, Waite T J, Boneberg E M, Woitsch S, Weiller M, Abela R, Grolimund D, Potin P, Butler A, Luther III G W, Kroneck P M H, Meyer-Klauck M and Feiters M C 2008 *Proceedings of the National Academy of Science* **105** 6954
[21] Palmer C J, Anders T L, Carpenter L J, Küpper F C and McFiggans G B 2005 *Env. Chem* **2** 282
[22] Zafiriou O C 1975 *Journal of Marine Research* **33** 75
[23] Carpenter L J, Wevill D J, Hopkins J R, Dunk R M, Jones C E, Hornsby K E and McQuaid J B 2007 *Geophys. Res. Lett.* **34** 1029
[24] Quack B and Wallace D W R 2003 *Global Biogeochem. Cy.* **17** 1023
[25] Moore R M and Zafiriou O C 1994 *Journal of Geophysical Research* **99** 16415
[26] Carpenter L J, Liss P S and Penkett S A 2003 *Journal of Geophysical Research* **108** 4256
[27] Palmer C J and Reason C J 2009 *Global Biogeochemical Cycles* **23** GB2014
[28] Carpenter L J 2003 *Chemical Reviews* **103**(12) 4953
[29] Ekdahl A, Pedersen M and K Abrahamsson 1998 *Marine Chemistry* **63** 1
[30] Wevill D J 2005 Ph. D. Thesis, *University of York, UK*
[31] Shall C, Heumann K G and Kirst G O B *Fresenius Journal of Analytical Chemistry* **359** 298
[32] Quack B, Atlas E, Petrick G, Stroud V, Schauffler S and Wallace D W R 2004 *Geophys. Res. Lett.* **31** 1029
[33] Smythe-Wright D, Boswell S M, Breithaupt P, Davidson R D, Dimmer C H and Eiras Diaz1 L B 2006 *Global Biogeochemical Cycles* **20** GB3003
[34] Yokouchi Y, Nojiri Y, Barrie L A, Toom-Sauntry D and Fujinuma Y 2001 *Journal of Geophysical Research* **106** 12661

[35] Moore R M and Tokarczyk R 1993 *Global Biogeochemical Cycles* **7** 195
[36] Class T H, Kohnle R and Ballschmitter K 1986 *Chemosphere* **15** 429
[37] Atlas E, Pollock W, Greenberg J and Heidt L 1993 *Journal of Geophysical Research* **98** 16933
[38] Moore R M, Webb M, Tokarczyk R and Wever R 1996 *Journal of Geophysical Research* **101** 20899
[39] Hopkins F 2009 *3rd International SOLAS Science Conference*, Barcelona pp. 41
[40] Liss P S 2009 *3rd International SOLAS Science Conference*, Barcelona, In perr. comm.
[41] Palmer C J and Reason C J *3rd International SOLAS Science Conference*, Barcelona pp.36

Several PBL parameterization lessons arrived at running an NWP model

Fedor Mesinger

Earth System Science Interdisciplinary Center (ESSIC), University of Maryland, College Park, MD 20740-3823, USA

E-mail: fedor@essic.umd.edu

Abstract. Several planetary boundary layer and turbulence transport parameterization advancements are reported upon, with the majority of them discovered by looking for a solution to a problem noticed in the course of running an operational NWP model. A surprising degree of sensitivity was repeatedly discovered to solutions found, with the refinements resulting in substantial improvements in the realism of forecasting a variety of actual atmospheric events. These improvements came from a reformulation of the Mellor-Yamada level 2 parameterization of surface fluxes via a change in the associated length scale; from a number of refinements in the specification of the Mellor-Yamada level 2.5 master length scale; and from the discovery of the reason for and improved handling of its realizability problem. Examples of the impact are shown. A refinement is summarized of the explicit molecular sublayer parameterization of Janjic, used over water as a constant-gradient approximation of the Liu, Katsaros and Businger relations. The refinement is arrived at by application of a momentum flux data summary and a suggestion by Brutsaert.

1. Introduction

The boundary layer or turbulence transport features to be addressed in the sequel are somewhat disparate to the extent that instead in one they could have just as well been discussed in several papers. Yet, their common characteristic is that they all have been arrived at by efforts to have the performance of a single regional weather prediction model, usually referred to as the Eta model, improve its performance, in most cases so as to correct very visible problems noticed in its forecasting of day-to-day atmospheric events. In other words the motivation for improvements to be reported on came from following the performance of a single model run in real time. The intention of the present paper is not to provide a detailed description of the Eta parameterization schemes as currently used, but rather to point out major advances arrived at in efforts to understand reasons of performance problems, and in discovering solutions to problems encountered. Parameterizations in models tend to be considerably different and thus frequently rather specific to a particular model; yet, it is felt that some of the points to be highlighted are relevant for approaches of a scope well beyond that of the specific model at hand that led to their being arrived at. At the same time, it is expected that some of the points to be covered should be applicable to a number of other parameterization packages given that similarities exist across models in more than one of the topics to be covered.

A specific feature of the Eta model is its eta coordinate, resulting in approximately horizontal coordinate surfaces. This has implications for boundary layer modeling since with the eta it is not possible to use a thin atmospheric layer everywhere close to the ground as normally done with the traditional terrain-following or sigma coordinate. Instead, for computational economy, in the Eta thicker and thicker model layers are used next to the ground as elevation increases. This of course is a challenge for PBL modeling, but as somewhat of a reward it is precisely this feature that led to some of the discoveries to be reported upon here. The issues encountered are of course relevant for sigma models as well; it is only that with sigma the problems faced would have been less severe and thus harder to notice. But it should be noted that while unique at the beginning [1], in recent years the eta or eta-like coordinate seems to be the choice of an increasing number of modelers [2], presumably because of its advantage of the avoidance of pressure gradient force errors that accompany the terrain-following approaches to a greater and greater degree as the steepness of the coordinate surfaces increases [3]. Yet another feature of the Eta model relevant for most of the topics to follow is its layer structure in the sense that model dynamical variables refer to model box average but are considered valid at grid box mid-points in terms of pressure. Fluxes on the other hand are typically defined at layer or grid box interfaces. This grid box approach is perhaps nowadays generally seen as advantageous in particular for climate modeling, as it facilitates design of discretization schemes that conserve total values of quantities of physical importance such as various forms of energy, and enstrophy, as, e.g., done in [4].

The two features just listed, impossibility to have a customary thin model layer everywhere next to the ground surface, and the layer design of the model, led to discoveries related to the turbulence length scales used. One of those refers to the length scale of the Mellor-Yamada level 2 surface fluxes scheme originally chosen for the model, expecting that it might be advantageous to the usual Monin-Obukhov similarity for thick lowest layers [5]. Another refers to the definition and placement of the Mellor-Yamada level 2.5 length scale, in case layers of unequal thickness are used as it is normally done. While a few other choices were helpful also, the two just mentioned led to quite significant improvements in the realism of the forecasts at hand. This material will be presented in the following section. A chance event having to do with the quest for turbulence in the higher troposphere led to the realization of a glaring problem in the "textbook way" of addressing the Mellor-Yamada 2.5 realizability problem. This, and the obvious remedy seen as a result, will be the subject of section 4. Section 5 will very briefly summarize the explicit parameterization of the molecular sublayer of Janjic and the work in progress on its refinement via application of the momentum flux data and a suggestion by Brutsaert. The paper will end with a concluding comment. All of this will be presented in an essay or perhaps memoir-type fashion emphasizing the history of the events that led to the discoveries or refinements made, believing this to be helpful to the readers and/or conducive to further studies.

2. Surface layer turbulence length scale

At the time when the original Eta model physics package was put in place, [5], vertical resolution used at NCEP was not high and a rather thick bottom layer was used. While a modification of the Betts-Miller convection scheme was successful in removing a problem with a case of spurious hurricane-like precipitation over warm water [6], a problem was discovered later with clearly excessive surface fluxes over warm water in other cases. One situation when this happened is the so-called "Valentine case," with its then National Meteorological Center's (NMC) surface analysis, valid 0000 UTC 16 February 1991, shown in the left panel of figure 1. 48-h Eta model forecast of the sea level pressure and 1000-500 millibars (mb) thickness verifying at the same time, is shown in its right panel. An overdevelopment is seen along the leading trough off the U.S. East Coast, with an erroneous center about 10 mb deeper than observed, while the double-center low over land is not as deep as the analyzed 986 mb. The weaker secondary through along the East Coast is poorly if at all forecast.

The Mellor-Yamada level 2 (MY-2) scheme [7, 8] associated with this overdevelopment problem contained several assumptions [5]. One was that of a constant shear above a thin logarithmic layer next to the ground. The assumption identified as responsible for the excessive fluxes was that of the MY-2 length scale determining the fluxes in the lower half of the bottom layer, with a finite-difference type value of 1/4 of its value on top of the layer, where it was made equal to that of the frequently used Blackadar's scheme.

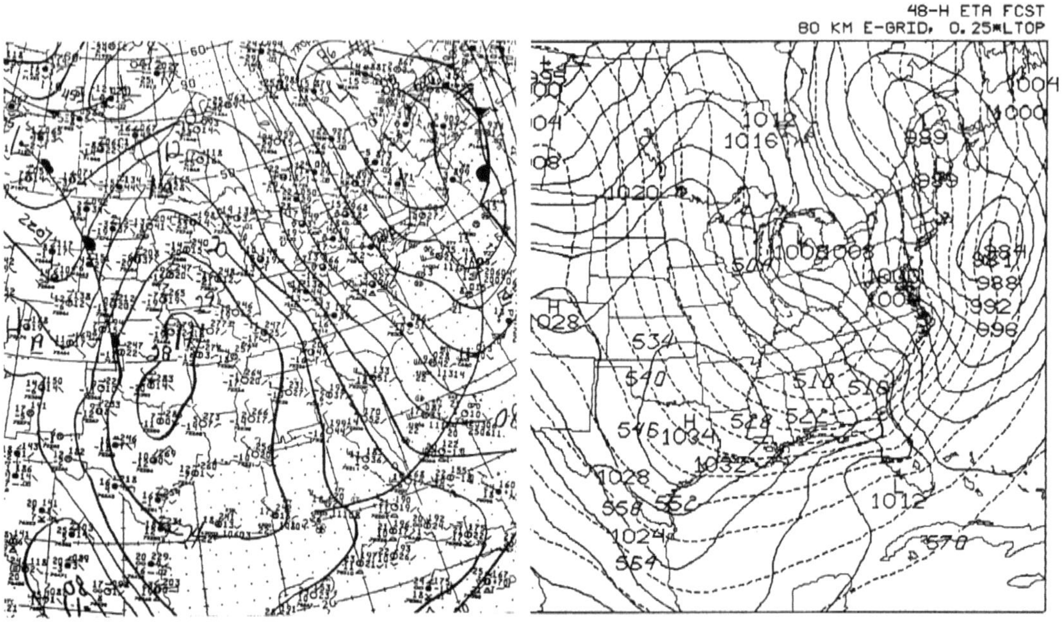

Figure 1. A section of the NMC surface analysis valid 0000 UTC 16 February 1991, left panel; 48-h Eta model forecast of the sea level pressure (full lines, mb) and 1000-500 mb thickness (dashed lines, decameters) verifying at the same time, right panel.

It has proven possible to replace these assumptions by an approach based on the standard approximation of a constant momentum flux in the lower half of the bottom layer,

$$\rho F_U = M = \text{const}, \tag{2.1}$$

with ρ standing for density, and U for the mean velocity. Inserting here the MY-2 expression for the momentum flux, [7], we have

$$\rho F_U = \rho l^2 \frac{dU}{dz}\left[B_1(1-R_f)S_M\right]^{1/2} S_M \frac{dU}{dz}, \tag{2.2}$$

with l being the length scale, z the height above the surface, B_1 a constant, R_f the flux Richardson number, and S_M the so-called (momentum) stability function. Replacing in the above l by its Blackadar length scale relation,

$$l = l_0 kz/(kz + l_0), \quad l_0 = \alpha \int zqdp / \int qdp, \quad \alpha = \text{const} \tag{2.3}$$

assumed now to be valid throughout the lower half of the bottom layer, we obtain

$$\left(\frac{kz}{kz+l_0}\frac{dU}{dz}\right)^2 = M\left(\rho l_0^{\,2}\left[B_1\left(1-R_f\right)S_M\right]^{1/2} S_M\right)^{-1}, \tag{2.4}$$

with l_0 being the Blackadar relation's asymptotic length scale, k the von Kármán constant, $q^2/2$ the turbulence kinetic energy, and p pressure. An upper limit of about 80 m was imposed on l_0. Solving the differential equation (2.4) is more complex than it may appear, since R_f depends on the sensible heat flux as well as on dU/dz, and the S_M, in addition, is itself a function of R_f. However, assuming the right side of (2.4) to be a constant seems like a reasonable first approximation, given that its expression multiplying M indeed does not depend on R_f in the neutral case. One then obtains

$$U = C\left(\frac{l_0}{k}\ln z + z\right) + D, \tag{2.5}$$

where C is the square root of the right side of (2.4), and D the integration constant. A well-known log-linear profile is seen to have followed from the assumptions made, suggestive of the stability effects having been retained.

The two constants in (2.5) can be chosen to have U satisfy the two values that are known at each end of the model's surface layer,

$$U(z_{Lm}) = U_{Lm}, \quad U(z_{1u}) = U_1, \tag{2.6}$$

with z_{Lm} and z_{1u} denoting the height of the mid-point of the lowest model layer, and that of the (momentum) molecular sublayer of the model, respectfully. Recall that in the Eta molecular sublayer over water is explicitly parameterized; thus, values on top of the sublayer are different from those at the surface. Inserting (2.6) into (2.5) leads to

$$\frac{U - U_1}{U_{Lm} - U_1} = \frac{(l_0/k)\ln(z/z_{1u}) + z - z_{1u}}{(l_0/k)\ln(z_{Lm}/z_{1u}) + z_{Lm} - z_{1u}}. \tag{2.7}$$

This, along with (2.3), can now be used in (2.2) to calculate the momentum flux. Note that, according to (2.4) and the assumption of its right side being constant, $l^2(dU/dz)^2$ is independent of height while the factor they are multiplied by in (2.2) is assumed constant so that this can be done at any height in between and including z_{1u} and z_{Lm}. For convenience though, and in the spirit of "bulk" relationships, one can define a "bulk" length scale, l_b, by

$$l^2\left(\frac{dU}{dz}\right)^2 \equiv l_b^{\,2}\left(\frac{U_{Lm} - U_1}{z_{Lm} - z_{1u}}\right)^2. \tag{2.8}$$

Inserting here (2.3) and (2.7), we obtain

$$l_b = l_0 / \left[1 + (l_0/k)\ln(z_{Lm}/z_{1u})/(z_{Lm} - z_{1u})\right]. \tag{2.9}$$

Thus, within the approximations made, the "l bulk" above is arrived at as the appropriate value to use in (2.1) along with a finite-difference gradient of U to calculate the momentum flux.

Analogous procedures can be followed using expressions for the sensible and latent heat fluxes; they lead to results same as (2.9), except that the corresponding heights of the temperature and moisture sublayers, z_{1q} and z_{1q}, take the place of z_{1u}. In replacing the Eta surface fluxes schemes this difference was disregarded and the same momentum-derived value of l bulk was used for all three fluxes.

The result obtained in our Valentine case with the "l bulk scheme" replacing the "0.25 l top scheme" is shown in figure 2. The overdevelopment off the East Coast is very much removed, so that it now shows as only a secondary development off the main low over land as it should. Both the leading and the secondary trough can be discerned, with even some mesoscale ridging in between them. The forecast overall might be considered as quite successful given that it has been obtained using the modest horizontal and vertical resolutions of the time, of 80 km and 16 layers, with the lowermost layer over water as much as 272 m thick. Note that while such resolutions in NWP nowadays would be well below the state-of-the-art, in climate modeling resolutions of that order may still be needed if very long integrations are to be done.

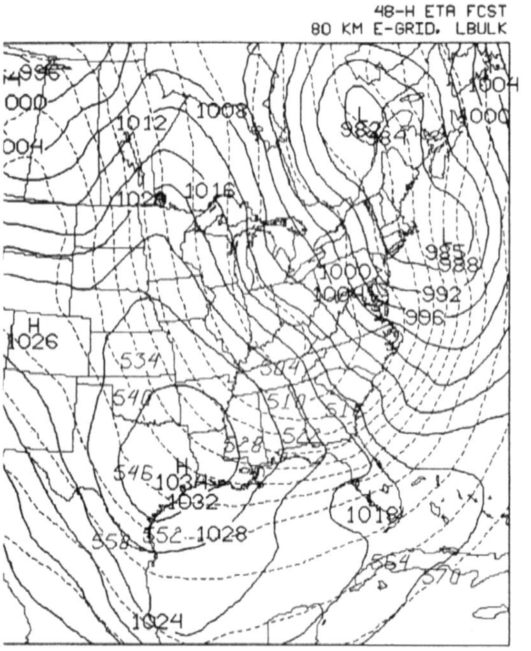

Figure 2. Same as the right panel of figure 1, except for the "*l* bulk" surface fluxes scheme replacing the "0.25 *l* top" scheme.

3. Mellor-Yamada 2.5 master length scale

While the *l* bulk scheme of the preceding section improved the Valentine case forecast greatly compared to the right side of figure 1, in relative terms the center off the East Coast is still slightly deeper than analyzed. There is also a center forecast over south Florida, which was not analyzed. In Fig 2 it is not as deep as in Fig 1, but it is not removed.

During the first winter of the real-time running of the scheme there were occasionally other signs suggestive of still somewhat too intense heating over the warm Gulf Stream water. While removal or refinements of some of the assumptions of the scheme are possible they are not straightforward, and as a simpler alternative it was decided to increase the vertical resolution near the ocean surface. In view of the modest computing resources of the time, this was done by splitting the model's lowest layer into two layers, one of 252 and the other of 20 m. Disappointingly, this led to a visible deterioration of the forecast shown in figure 2. The development over water off the East Coast had almost completely disappeared, and no sign of the intense leading trough was any more visible.

Looking for a possible cause of this unwelcome development, attention was focused on the specification of the turbulence length scale. Recall, [5], that it is the Mellor-Yamada 2.5 scheme (MY-2.5) that was and is used in the Eta for turbulence exchange above the surface layer; thus, it is the MY-2.5 master length scale that we are considering now. With MY-2.5, the exchange coefficients are proportional to the master length scale, at the time evaluated at layer interfaces [5, figure 3]. This clearly seemed like a suspect feature, with the shallow lowest layer and a much thicker layer above. Note that defined by (2.3) the master length scale will be small when the top of the lowest layer is low, consistent with the disappearance of the intensive development over water.

The Eta numerical design being one of a "layer" model makes it only natural to look at the master length scale also as a characteristic of the model layer, and to use for the exchange of properties between neighboring layers an average of the two layer values. Thus, calculation of the master length

IOP Conf. Series: Earth and Environmental Science **13** (2010) 012005 doi:10.1088/1755-1315/13/1/012005

scale at model interface elevations was replaced by the calculation at layer mid-points, and two-layer averaging to obtain the interface values,

$$l_{L+1/2} = \frac{1}{2}\left(l_L + l_{L+1}\right), \tag{3.1}$$

the subscript L here being the model layer index, increasing downward.

The impact of the change from the interface values to the "averaged l" values (3.1) is shown in figure 3. In its left panel is the Valentine case 48-h forecast obtained using (2.3) and the 17-layer model, with hardly any evidence left of the leading trough off the U.S. East Coast once the thin 20 m layer has been added. In its right panel is the same forecast except for using (3.1). Development off the East Coast seen in the analysis shown in figure 1 has been recaptured to a considerable degree, with a clear indication of the position of the trough now seen in the shape of the isobars of 996 and 1000 mb. In association with this the overdeepening of the center over Quebec has been reduced to a value closer to the one analyzed.

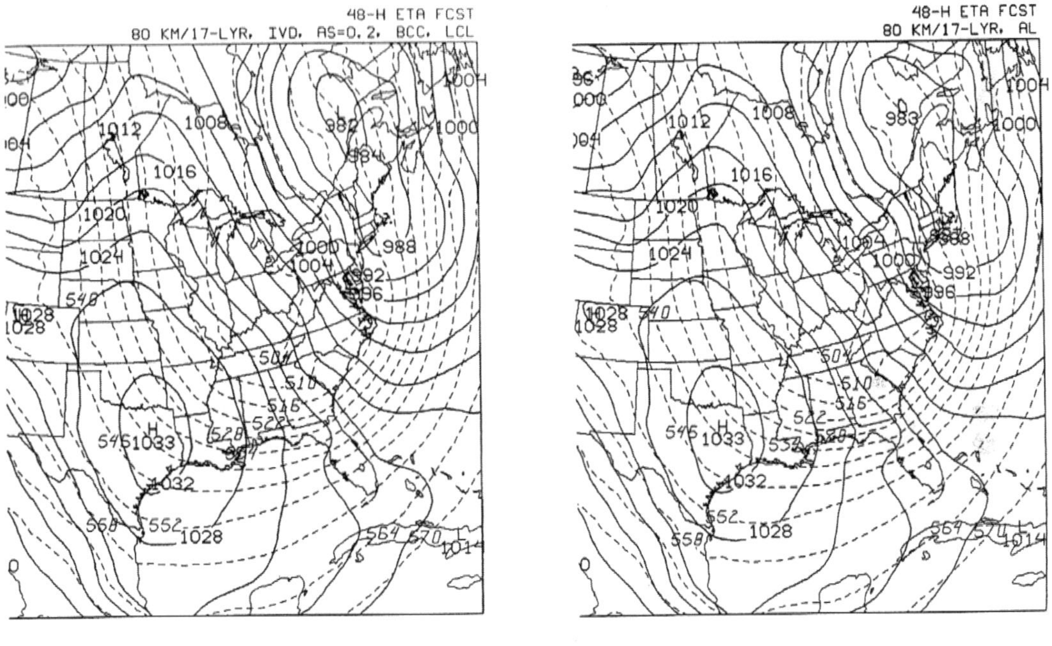

Figure 3. Same as figure 2, except for the lowest model layer being split into two layers, with the bottom layer of 20 m (left panel); same as the left panel, except for the "averaged l" scheme (3.1) used to calculate the interface master length scales, right panel.

Several more changes have been introduced for a more realistic specification of l, given that the simple Blackadar relation (2.3) was clearly not appropriate for a model extending well beyond the PBL into the troposphere and higher. One was the frequently used [e.g., 9] stability restriction according to the dimensional formula of Deardorff [10] with l not being allowed to exceed

$$l_D = c_r q / N, \tag{3.2}$$

c_r here being a constant, and N the Brunt-Väisälä frequency. Another was the restriction of the Blackadar scheme to the boundary layer, at the time assuming for that purpose the boundary layer depth z_i to be proportional to the Blackadar's asymptotic length scale,

$$z_i = c_g l_0, \tag{3.3}$$

c_g being another proportionality constant. The simple relation (3.3) was previously used in [11] and [12]. For layers below z_i we have defined l as

$$l = \min(l_B, l_D) \tag{3.4}$$

where l_B stands for the Blackadar value (2.3). Above the boundary layer, we have defined l as

$$l = \min(l_\Delta, l_D, kz), \tag{3.5}$$

where

$$l_\Delta = c_s \Delta z, \tag{3.6}$$

is a prescribed fraction of the vertical grid distance, Δz, [13; see also 14, and 15]. For the proportionality constants c_r, c_g, and c_s originally the values of 0.25 [16], 10, and 0.23 (Lilly's inertial subrange value) were used, respectively.

The forecast obtained with the length scale scheme (3.2) through (3.6) and constants as stated is shown in figure 4. The definition of the position of the trough off the U.S. East Coast has further improved compared to that of the right panel of figure 3. Note the increased sharpness of the kinks in the isobars of 992 and 996 mb off the coast, with a better indication of the mesoscale ridge separating this leading trough from the second one running toward the Carolinas. This ridge can be seen as a conspicuous feature of the verification analysis, left panel of figure 1.

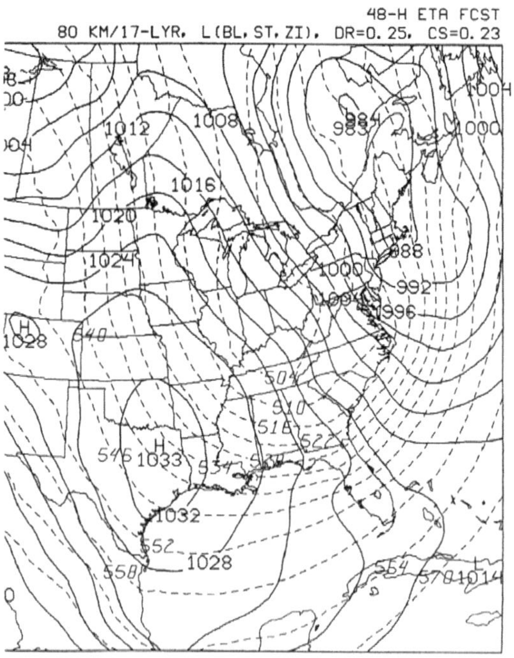

Figure 4. Same as the right panel of figure 3, except for using the scheme (3.2) through (3.6) to calculate layer values of the master length scale, instead of (2.3).

Somewhat of a replay of the thicknesses story that led to (3.1) but with eventually a still stronger message occurred much later when considerably higher vertical resolutions and thus thinner layers were operationally affordable, and lowest layers thicknesses were used not nearly as lopsided as that of the 252 next to 20 m. During experimental running of the 32 km/45 layer Eta in 1997 it was noted that this experimental model version failed to forecast the development of Hurricane Danny over the Gulf waters as coherently and as intensely as the operational Eta did, in spite of the lower resolution of the operational Eta, of 48 km/38 layers. This was found worrisome as improved performance was of course expected from higher resolution and a search for a possible cause was undertaken.

A suspect feature was identified in the experimental Eta's next to the lowest layer thickness, of 58 m, being greater than that of the operational Eta of 43 m, the situation thus being reminiscent of the problem with the 17 layer model described above. This led to an eventual discovery of the layer averaging scheme (3.1) having been inadvertently omitted when rewriting the length scale subroutine of the code. With the omission being rectified in the experimental 32 km/45 layer code but not yet in the operational code, inspection of 24-h forecasts of a vertical sounding at Key West, FL, by the two codes, revealed a dramatic difference between them, as shown in figure 5. The code not using the averaging scheme, long/ short dashes in the figure, failed to forecast the inverted V marine boundary layer dew point and temperature profiles as observed, shown by the solid lines. Instead, moisture is seen to be restricted to a very shallow layer capped by dry layers above. The code with the averaging scheme restored, shown by the dashed lines, is seen to have produced a realistic boundary layer forecast with the mixed layer height agreeing well with that of the sounding, supporting thus the diagnoses of the reason for the Hurricane Danny problem referred to above.

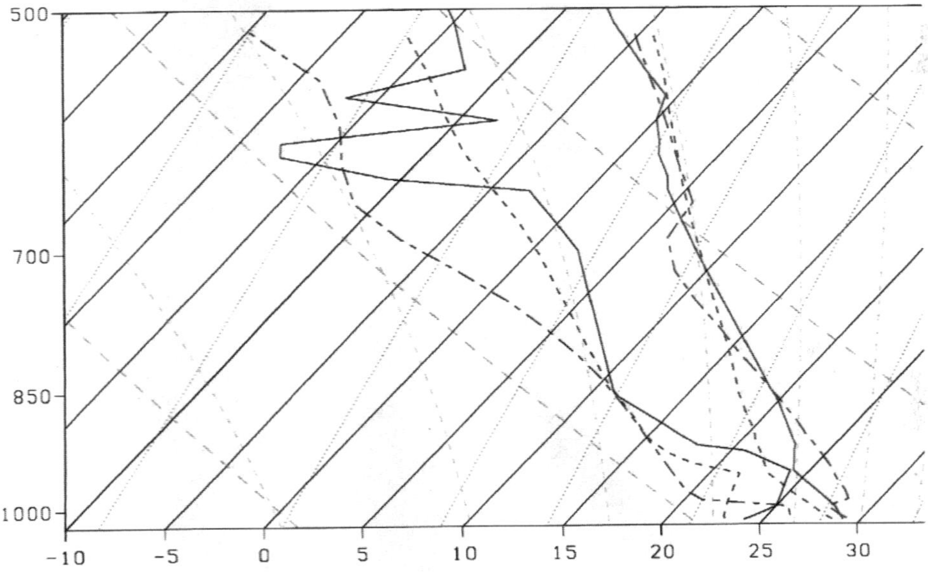

Figure 5. Observed dew point and temperature soundings (solid lines), and same except 24-h forecasts by two versions of the Eta model, one which did not use the length scale averaging scheme (3.1), long/short dashes; and another which did, short dashes; all verifying 0000 UTC 31 July 1997. Temperature, in degrees C, is on the abscissa, and pressure, in mb, on the ordinate; due point lines are to the left of the corresponding temperature lines. (Joe Gerrity, personal communication.)

4. Mellor-Yamada level 2.5 realizability problem

A puzzling situation arose as a result of the initial efforts of the then NOAA Forecast Systems Laboratory (FSL) group [17] to check if explicit Eta model forecasts of upper tropospheric TKE might be useful for operational forecasting of clear air turbulence. Looking at a few simulations done at FSL, of interest as there were pilot reports of turbulence at the time, no upper-tropospheric TKE worth any mention was noticed. This being worrisome, upper tropospheric TKE plots for a total of four 48-h Eta 80-km/17 layer integrations were produced at NMC and inspected. In three out of these four cases substantial amounts of upper tropospheric TKE were noted. Turbulence speed maxima were as high as more than 1 to 3 m s^{-1}.

One difference between the model as used for the simulations done at FSL and that used for the four cases subsequently run at NMC was that the latter contained the stability dependent specification of the master length scale (3.2) instead of the earlier boundary-layer type Blackadar scheme. But that seemed like a poor candidate to explain the difference in the behavior of the two models: note that both the shear and the buoyancy TKE production terms [e.g., 5]

$$P_s = \frac{q^3}{l} S_M G_M, \quad P_b = \frac{q^3}{l} S_H G_H, \tag{4.1}$$

in view of

$$G_M \equiv \left(\frac{l}{q}\right)^2 \left[\left(\frac{\partial U}{\partial z}\right)^2 + \left(\frac{\partial V}{\partial z}\right)^2\right], \quad G_H \equiv -\left(\frac{l}{q}\right)^2 \beta g \frac{\partial \Theta_v}{\partial z}, \tag{4.2}$$

β being the coefficient of thermal expansion, g gravity, and Θ_v the mean flow virtual potential temperature, are proportional to l^2, and that the stability dependent l scheme was in the upper troposphere giving values of l typically very much smaller than the Blackadar scheme. Even so, when one of the TKE endowed cases was rerun at NMC with the Blackadar scheme, instead of the substantial upper tropospheric TKE which at one of the output times previously was present, negligible TKE values were obtained.

The solution to this puzzle was found in the clipping imposed on G_H and G_M [5, see also 8]

$$G_H \le 0.024, \quad G_M \le 0.36 - 15 G_H. \tag{4.3}$$

Since the first restriction (4.3) is imposed only when G_H is positive, and enforcement of either one or both of (4.3) leaves G_H as well as G_M non-negative, one can see that as far as the values of G_M and G_H are concerned an imposition of (4.3) is the same as if one had in their definition (4.2) reduced the value of l. Yet, in the final evaluation of the sum of the production and dissipation terms,

$$P_s + P_b - \varepsilon = \frac{q^3}{l}\left(S_M G_M + S_H G_H - \frac{1}{B_1}\right), \tag{4.4}$$

ε being the dissipation, and B_1 one of the M-Y empirical constants, the original unreduced l was used. It was clearly possible that in this way a relatively large value of l could not only lead to physically unjustified imposition of (4.3) just when the model was trying to generate the TKE, but that in addition the division by this large value of l in (4.4) would make the impact of these reduced values of G_M and/or G_H still less effective, to the extent of taking away altogether the model's ability to generate a meaningful amount of the upper tropospheric TKE.

For a test of this hypotheses an experiment was run in which (4.3) was not imposed by clipping the values of G_H and G_M but instead by a reduction of l. Use of the Blackadar specification with its tropospheric values of l very much larger than those resulting from (3.5) did indeed in this test result in the generation of significant amounts of upper tropospheric TKE, as shown in the left panel of figure 6. Interestingly, the TKE pattern obtained is quite similar to that resulting from the stability dependent l scheme (3.5), shown in the right panel of the figure. For additional comfort one might note that the maximum amounts obtained with the large l values of the left panel are now greater than those obtained with the smaller values of the right panel, as reasonable on physical grounds.

It would seem that the enforcement of the realizability constraint in the form of (4.3) and clipping of G_H and G_M instead of l represented a major oversight on the part of previous authors looking into this issue [8, 18, 5]; and that it is likely to have at times and places detrimentally affected the performance of the scheme, as it did in our case. Note that the mechanism of the impact of (4.3) as outlined is not restricted to troposphere so that it can occur in the PBL at times when generation of TKE is physically justified just as well.

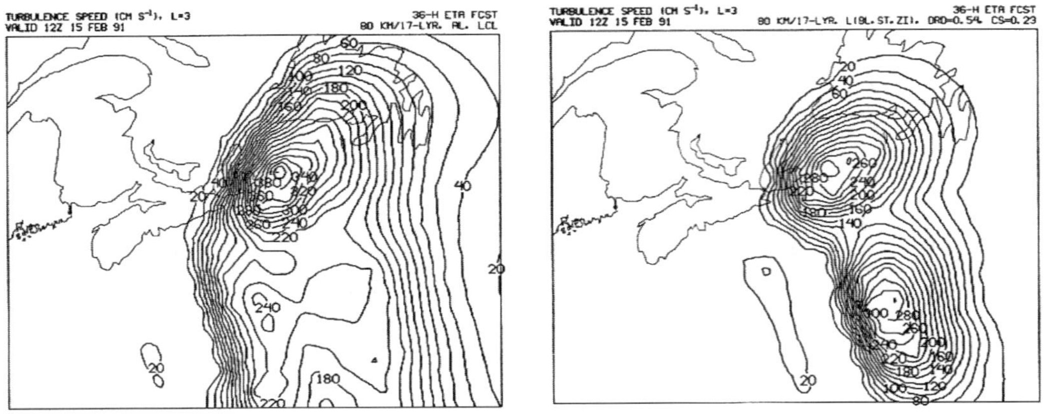

Figure 6. Turbulence speed (cm s^{-1}) at the interface of layers 3 and 4 (about 250 mb) of the Eta 80-km/17 layer model, obtained in a 36-h forecast valid 1200 UTC 15 February 1991, using the Blackadar scheme (2.3) for the prescription of l and clipping of l to enforce the realizability condition (4.3), left panel; same except for the stability dependent scheme (3.5) for the prescription of l, right panel. The sections of the model domain shown encompass 10 deg of latitude from their southern to their northern boundaries.

On the more positive side, given that the specification or prediction of the length scale is possibly the most uncertain territory of the Mellor-Yamada model [e.g., among others, 8, 19] we find it reassuring that in the experiment shown two schemes which before the turbulence event were giving drastically different values of l still resulted in the generation of turbulence at about the same time and place and with patterns as similar as those shown in our figure 6.

5 Work in progress: molecular sublayer depth refinement

A nonstandard feature of the Eta PBL parameterizations is the explicit parameterization of the molecular sublayer over water [6]. The parameterization is a constant-gradient approximation of the relations of [20, LKB later on], with parameters based on the experimental data of [21]. In this parameterization the exponential smooth regime relationships of LKB with their gradual transition into the turbulent surface layer are replaced by molecular sublayers in which transports are determined entirely by molecular diffusion, matched to log profiles above in which transports are entirely turbulent. In addition, for chosen values of friction velocity, the sublayers for the three variables handled undergo discontinuous regime changes, from smooth to rough, and into "rough with spray", as suggested by the data of [21]. When switching from smooth to the rough regime, the sublayer for momentum is removed, while the thicknesses of the sublayers for the sensible and the latent heat fluxes are reduced to one third of their smooth regime values. Switching into the rough with spray regime, these two sublayers are removed as well.

Subsequent to the implementation of the scheme taking advantage of an exercise in progress involving typhoon Yancy (Eric Rogers, 1992, personal communication) an experiment was performed in which two simulations were performed by the Eta model of the time, one with and one without the molecular sublayer parameterization. The motivation was that a typhoon being fed by evaporation could well be an event in which the explicit parameterization of molecular sublayer could be make a difference. Initial time was chosen so as to have Yancy attain its maximum wind strength at the 60-h simulation time. The model was run using a 50 km/20 layers resolution. The run without the molecular sublayer ended with Yancy of insufficient strength and with an elongated shape not typical of typhoons, while another low some 1500 km east was deeper than it ought to have been. With the molecular sublayer, both of these problems were indeed dramatically improved (not shown; see the plots at http://crg.wits.ac.za/ct2009/Session4/Mesinger_CT2009a.pdf, pp 29–30).

It was later noted possible to put what was considered to be a free parameter of the scheme on a firmer footing. Namely, a key parameter of the scheme, ζ, was in [6] defined by postulating

$$\frac{z_{1u}u_*}{C\nu} = \frac{z_{1\theta}u_*}{S\kappa} = \frac{z_{1q}u_*}{D\varepsilon} = \zeta, \tag{5.1}$$

with z_{1u}, $z_{1\theta}$, and z_{1q} being sublayer thicknesses for momentum, temperature, and moisture, respectively, u_* the friction velocity, C, S, and D parameters known as the inverse interfacial drag coefficient, and inverse interfacial Stanton and Dalton numbers, respectively, and ν, κ, and ε the kinematic viscosity, thermal diffusivity, and molecular diffusivity of water vapor, respectively. ζ was considered to be a tuning parameter, with the value of 0.50 used in [6].

One can however ask a question: if the linear profile at the bottom of the molecular sublayer is linearly extrapolated upwards, and the logarithmic profile of the surface layer is at the same time logarithmically extrapolated downwards, at what elevation will the two extrapolated profiles intersect? This should be the appropriate value of z_{1u}, from which ζ can be calculated; and is in fact a procedure that was suggested by Brutsaert [22]. Brutsaert also gives the result for the nondimensional height used in his diagram showing median of extensive experiments. Using his value, simple algebra leads to

$$\zeta = 11/G\mathrm{Rr}^{1/4}, \tag{5.2}$$

where G is a proportionality constant, and

$$\mathrm{Rr} = z_0 u_*/\nu$$

is the roughness Reynolds number, z_0 being the roughness height. The widely used Charnock relation is used for z_0, and the "Charnock constant" of 0.018. Fitting laboratory data to their profile relationships LKB obtain for the smooth regime the value of G close to 30, used in [6].

The value (5.2) is used in the "upgraded version of the Eta" (http://www.cptec.inpe.br/etaweb/). Experiments testing its impact compared to a constant are in progress, with preliminary results shown also at http://crg.wits.ac.za/ct2009/Session4/Mesinger_CT2009a.pdf (Josiane Bustamante, personal communication). While changes obtained are relatively small, they do appear to be physically based with some of the ITCZs visibly shifted in meridional directions within several days of the integrations done.

6 Concluding comment

For PBL research a non-standard feature of the results and in some cases discoveries presented is that they have all but one been arrived at by striving to correct a non-acceptable performance of an NWP model. It is felt that this is not a weakness but rather a strength of the results presented, because, to quote from [23] the impact of "connection to the basic physics" as opposed to "representations of stress and thermodynamics fluxes through one or two empirical constants", is at times much greater than what one might expect, and this seems to be the case particularly with events of considerable significance. Operational NWP models, due to the daily scrutiny of their performance, are an excellent vehicle for this type of effort.

References

[1] Mesinger F 1984 A blocking technique for representation of mountains in atmospheric models *Riv. Meteorol. Aeronautica* **44** 195–202
[2] Marshall J, Adcroft A, Campin J-M, Hill C and White A 2004 Atmosphere–ocean modeling exploiting fluid isomorphisms *Mon. Wea. Rev.* **132** 2882–94
Steppeler J, Bitzer H W, Janjic Z, Schättler U, Prohl P, Gjertsen U, Torrisi L, Parfinievicz J, Avgoustoglou E and Damrath U 2006 Prediction of clouds and rain using a z-coordinate

nonhydrostatic model *Mon. Wea. Rev.* **134** 3625–43

Russell G L 2007 Step-mountain technique applied to an atmospheric C-grid model, or how to improve precipitation near mountains *Mon. Wea. Rev.* **135** 4060–76

Walko R L and Avissar R 2008 The ocean-land-atmosphere model (OLAM). Part I: Formulation and tests of the nonhydrostatic dynamic core *Mon. Wea. Rev.* **136** 4045–62

[3] Mesinger F 2004 The steepness limit to validity of approximation to pressure gradient force: Any signs of an impact? *20th Conf. on Weather Analysis and Forecasting/16th Conference on Numerical Weather Prediction*, paper P1.19, 14 pp, Combined Preprints CD-ROM, 84th AMS Annual Meeting, Seattle, WA

[4] Janjic Z I 1984 Nonlinear advection schemes and energy cascade on semi-staggered grids *Mon. Wea. Rev.* **112** 1234–45

[5] Janjic Z I 1990 The step-mountain coordinate: Physical package *Mon. Wea. Rev.* **118** 1429–43

[6] Janjic Z I 1994 The step-mountain eta coordinate model: Further developments of the convection, viscous sublayer, and turbulence closure schemes *Mon. Wea. Rev.* **122** 927–45

[7] Mellor G L and T Yamada 1974 A hierarchy of turbulence closure models for planetary boundary layers *J. Atmos. Sci.* **31** 1791–806

[8] Mellor G L and T Yamada 1982 Development of a turbulence closure model for geophysical fluid problems *Rev. Geophys. Space Phys.* **20** 851–75

[9] Galperin B, Kantha L H, Hassid S and Rosati A 1998 A quasi-equilibrium turbulent energy model for geophysical flows *J. Atmos. Sci.* **45** 55–62

[10] Deardorff J W 1976 Clear and cloud-capped mixed layers - their numerical simulation, structure and growth and parameterization *Seminars on the Treatment of the Boundary Layer in Numerical Weather Prediction* (Reading, UK: ECMWF) pp 234–84

[11] Gambo K 1978 Notes on the turbulence closure model for atmospheric boundary layers *J. Meteorol. Soc. Japan* **56** 466–80

[12] Yamada T 1979 PBL similarity profiles determined from a level-2 turbulence-closure model *Boundary-Layer Meteorol.* **17** 333–51

[13] Lilly D K 1962 On the numerical simulation of buoyant convection *Tellus* **14** 148–71

[14] Deardorff J W 1980 Stratocumulus-capped mixed layers derived from a three-dimensional model *Boundary-Layer Meteorol.* **18** 495–527

[15] Mason P J 1989 Large-eddy simulation of the convective atmospheric boundary layer *J. Atmos. Sci.* **46** 1492–516

[16] Duynkerke P G and Driedonks A G M 1987 A model for the turbulent structure of the stratocumulus-topped atmospheric boundary layer *J. Atmos. Sci.* **44** 43–64

[17] Marroquin A, Greenhut G K, Jamison B, Mahoney J L, Schwartz B and Girz C M I R 1992 Modeling and forecasting clear air turbulence using a mesoscale numerical model *10th Symp. on Turbulence and Diffusion* (Amer. Meteorol. Soc.) pp 346–9

[18] Helfand H M and Labraga J C 1988 Design of a nonsingular level 2.5 second-order closure model for the prediction of atmospheric turbulence *J. Atmos. Sci.* **45** 113–32

[19] Burchard H 2001 On the q^2l equation by Mellor and Yamada (1982) *J. Phys. Oceanogr.* **31** 1377–87

[20] Liu W T, Katsaros K B, Businger J B 1979 Bulk parameterization of air-sea exchanges of heat and water vapor including the molecular constraints at the interface *J. Atmos. Sci.* **36** 1722–35

[21] Mangarella P A, Chambers A J, Street R L, Hsu E Y 1973 Laboratory studies of evaporation and energy transfer through a wavy air-water interface *J. Phys. Oceanogr.* **3** 93–101

[22] Brutsaert W 1982 *Evaporation into the Atmosphere* (Dordrecht: Reidel)

[23] Fairall C W, Hare J E, Edson J B and McGillis W 2000 Parameterization and micrometeorological measurement of air-sea gas transfer *Boundary-Layer Meteorol.* **96** 63–105

IOP Conf. Series: Earth and Environmental Science 13 (2010) 012006 doi:10.1088/1755-1315/13/1/012006

Temporal and spatial changes in mixed layer properties and atmospheric net heat flux in the Nordic Seas

A Smirnov[1,3], A Korablev[2], G Alekseev[1] and I Esau[2]

[1] SI "Arctic and Antarctic Research Institute", St. Petersburg, Russia
[2] Nansen Environmental and Remote Sensing Centre, Bergen, Norway
[3] Nansen International Environmental and Remote Sensing Centre, St. Petersburg, Russia

E-mail: avsmir@aari.nw.ru

Abstract. The Nordic Seas are an important area of the World Ocean where warm Atlantic waters penetrate far north forming the mild climate of Northern Europe. These waters represent the northern rim of the global thermohaline circulation. Estimates of the relationships between the net heat flux and mixed layer properties in the Nordic Seas are examined. Oceanographic data are derived from the Oceanographic Data Base (ODB) compiled in the Arctic and Antarctic Research Institute. Ocean weather ship 'Mike' (OWS) data are used to calculate radiative and turbulent components of the net heat flux. The net shortwave flux was calculated using a satellite albedo dataset and the EPA model. The net longwave flux was estimated by Southampton Oceanography Centre (SOC) method. Turbulent fluxes at the air-sea interface were calculated using the COARE 3.0 algorithm. The net heat flux was calculated by using oceanographic and meteorological data of the OWS 'Mike'. The mixed layer depth was estimated for the period since 2002 until 2009 by the 'Mike' data as well. A good correlation between these two parameters has been found. Sensible and latent heat fluxes controlled by surface air temperature/sea surface temperature gradient are the main contributors into net heat flux. Significant correlation was found between heat fluxes variations at the OWS 'Mike' location and sea ice export from the Arctic Ocean.

1. Introduction

The 'Nordic Seas' is a generalized name for the Norwegian, Greenland and Barents seas. This region is of primary importance for the climate of the Northern Europe. The Greenland–Scotland Ridge forms a barrier that constrains the exchange of water between the North Atlantic and the Nordic Seas. A warm, saline water flows northeastwards in the near-surface layers while cold, fresher water returns southwestwards partly in the near-surface layers along the Greenland coast and partly in the bottom layer at several locations across the ridge. This dens overflow results from cooling and subsequent sinking of near-surface layer waters in the areas north of the Greenland-Scotland Ridge. After passing the ridge the overflow water changes its properties considerably due to entrainment of surrounding Atlantic water turn into the North Atlantic Deep Water mass [8]. Oncoming advection of heat and salt plays vital role in shaping global thermohaline circulation and regional climate. The properties of the source water masses advected into the region are considerably modified along pathways by strong air-sea interaction, vertical and lateral mixing, especially during winter season.

Previous studies have shown that the vertical convective exchange in the Nordic Seas is highly variable and is very sensitive to the climate change [7,9,10].

The Nordic Seas are strongly affected by the air-sea exchanges of heat, salt and moment. Due to the warm Norwegian Current the atmosphere gets considerable amount of heat from the ocean. During the winter cooling a deep homogeneous layer is generated. Under favorable conditions the mixed layer can reach the bottom. The heat budget of the mixed layer is of particular interest because it governs the evolution of the sea surface temperature, which is the most important ocean parameter influencing the atmosphere.

The study of mechanisms of transfer of a climate signal from the surface to deeper layers is essential for understanding climate change. The intensity of the mixing processes depends on different factors, including stratification of the water column and atmospheric forcing parameters (air temperature, wind stress, humidity etc.). The article is focused on algorithms for quantitative estimations of mixed layer properties and air-sea fluxes.

2. Data

Oceanographic observations used in this paper were extracted from database compiled for the Nordic Seas under the INTAS-4620 project 'The Nordic Seas in the Global Climate System' [12,13] which was considerably updated recently. About 2.5 million of oceanographic stations from more than 30 original sources were processed. The quality control (QC) procedures were designed to eliminate low accuracy and biased data. For this purpose, only data measured by high quality instruments (bottle, CTD) were accepted for analysis. The final merged, quality/duplicate controlled database for 60°-82°N, 45°W-70°E geographic area contains more than 400,000 oceanographic stations with temperature, salinity, oxygen and chemistry profiles for 1900-2009.

In addition, a new surface layer dataset has been developed. It contains more then 550 000 quality and duplicate controlled stations with temperature and salinity within upper 5 m. Here, we did not apply any instrumental limitations on data because they are not so critical for the upper layer due to large natural variability.

Spatially uneven oceanographic observations were objectively analyzed to generate horizontal gridded fields. For this purpose, temperature and salinity profiles were interpolated at standard depth levels and integrated monthly. All observed monthly fields with measurements for 1900-2009 were converted to gridded fields with 0.25° x 0.25° resolution by means of Intrinsic Kriging (IK). IK model was selected because it allows to process non-stationary data with trends assessment inherent in oceanographic data. Derived monthly gridded fields were further used for mean (climatological) fields computing by means of arithmetic averaging for certain periods. Combined analysis of the climatological and monthly gridded field make possible to study long-term variability in the Nordic Seas and estimate anomalies in each particular year. It is also provide an input for quantification of mixed layer properties and air-sea interaction.

Ocean weather ship "Mike" located at 66°N, 2°E is the only fixed station in the Nordic Seas which has all required set of variables for the net heat flux calculations. It is the longest uniform and continuous dataset in the region commenced in 1948. Oceanographic profiles and meteorological parameters are taken synchronously at the same location that considerably reduces uncertainty derived from variables spatial interpolation.

3. Air-sea heat fluxes

There are many limitations in estimating of air-sea heat fluxes based on ship borne data. We decided to use following methods which correspond to the requirements. *Shortwave* incoming radiation was calculated by using Reed's formula [17]. An albedo data were extracted from satellite array constructed by Taskanen and described in [20]. At clear sky incoming radiation parameterization the EPA model [1] was utilized. *Longwave* radiation was parameterized by the Southampton Oceanography Centre formula for mid-high latitudes [11]. It was developed and tested especially for

mid latitudes. *Turbulent* heat fluxes were calculated by using COARE (Coupled Ocean-Atmosphere Response Experiment, 1990-2003) algorithm [5,6]. Initially that algorithm was developed by using COARE measurements, albeit with a relatively limited wind speed range (0–12 m/s), and has been used by many research groups for heat flux analyses (more than 200 citations) [5]. However, it has also been used outside the Tropics in a wide variety of conditions, raising obvious questions about its applicability in colder waters, mid latitudes, and much higher wind speeds. The algorithm was 'globalized' in the sense that, wherever possible, physical variables are computed as functions of ambient conditions (e.g., temperature, latitude, solar heat flux) rather than hardwired to tropical values. The recent changes and claimed preciseness have made COARE one of the best bulk-based heat flux algorithms to be used in the mid-high latitudes [6].
The net heat flux calculated from OWS "Mike" data is shown on Figure 1. There are three distinct pikes. The first occurred at the end of 60^s, the second in the beginning of 80^s and the last one appeared in the mid 90^s.

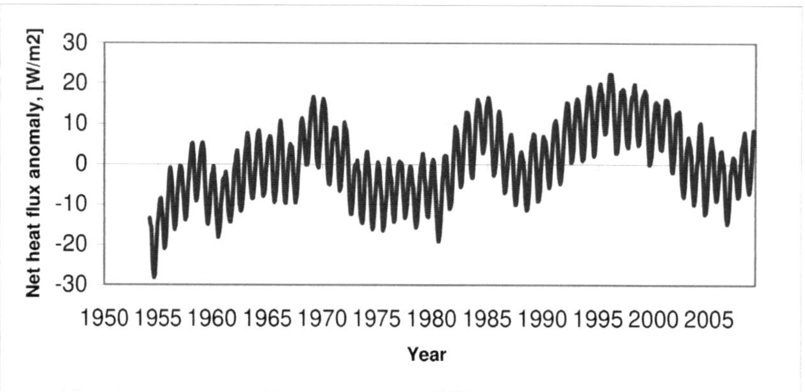

Figure 1. Averaged net heat flux anomaly time series calculated from OWS 'Mike' data

The biggest contribution into the total heat flux has been made by the sensible and latent heat fluxes (f

Figure 2). It means that during the indicated periods there were some processes which sharpen temperature gradients between the ocean and the atmosphere. Possible causes of sharp temperature changes will be discussed in **section 5**.

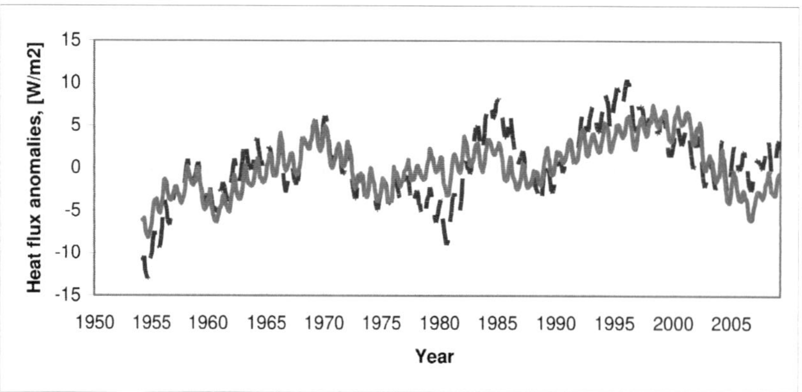

Figure 2. Averaged sensible heat flux anomaly (blue, dotted line) and latent heat flux anomaly (solid, red line)

4. Mixed layer depth

Oceanographic mixed layer (ML) is not as well investigated and sampled as the atmospheric boundary layer (ABL). There are some simple explanations of this discrepancy. First reason is sparse coverage and low vertical/temporal of the oceanographic variables required for the ML appropriate definition. Second, the mixed layer depth (MLD) definition is not straightforward and different algorithms give potentially dissimilar results. There are three main groups of methods:

1. Threshold methods (Finite difference criterions)
2. Gradient methods
3. Curvature of profiles

Other algorithms such as 'Step-function least squares regression' method [14], 'Integral depth-scale' method, 'Trapping' method [16], 'Split and merge' [15] are not being widely used because of their complexity of automation, auxiliary information and expert presence requirements [21].

Historical oceanographic data quite often have low vertical resolution and contain errors. The only applicable groups of algorithms are the 'Threshold' methods. But the results may have very big errors (up to 50 m) because of the algorithm and data limitations. The MLD for the OWS 'Mike' for the entire set of oceanographic data has been calculated by using of all listed above methods, but the results were quite far from the reality. We should note that Mike's time series have been sampled by bottles before 2002, with low vertical resolution. After 2002 year profiles are getting obtained by the CTD sounding. We repeated calculations for the period and correlated the results with the net heat flux (Figure 3). The correlation coefficient was equal *0.7-0.8* depending on algorithm of MLD definition. Good correspondence clearly shows that for correct definition of the ML properties high quality data are essential.

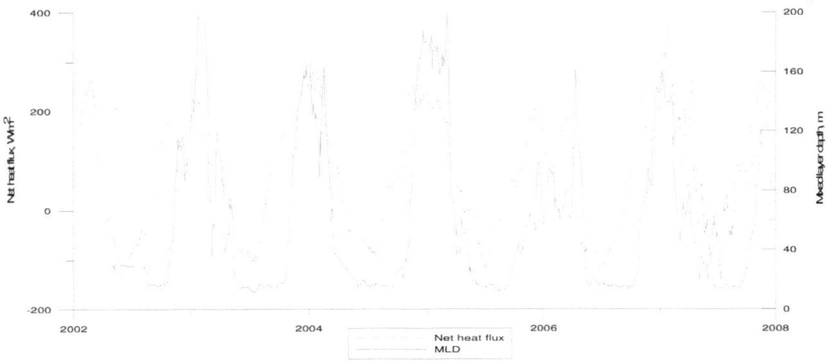

Figure 3. Inter comparison of the MLD and net heat flux

5. Discussion

It is well known that during the period of instrumental observations ocean properties in the Nordic Seas undergo several stages in hydrographic regime. Periods with stable regime were alternated with abrupt changes. To estimate temporal and spatial variability of water properties we used gridded monthly climatology briefly described in the **section 2**.

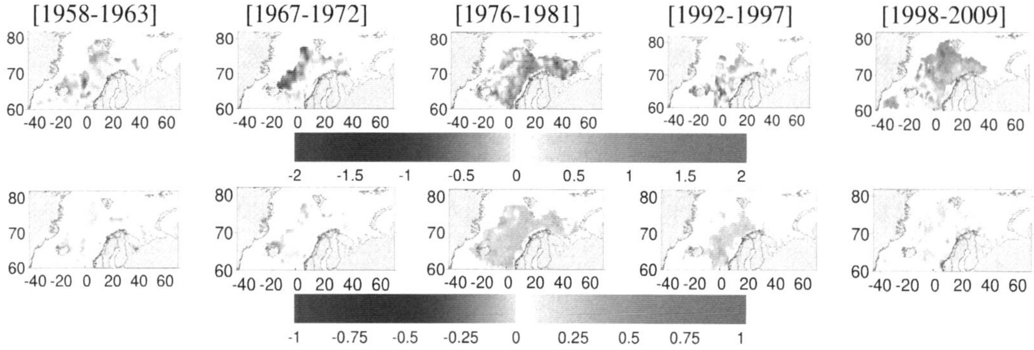

Figure 4. Spatial distribution of temperature (a) and salinity (b) at 50 m in April

Figure 4 shows the spatial distribution of anomalies of temperature and salinity for different periods with stable regimes relative to 1957-1990 mean. The late 50^s – early 60^s period was characterized by warmer and saltier conditions. At the end of 60^s the sea surface temperature (SST) was reduced while salinity background remained positive. It favored ('preconditioning') the deep convection development. Response of the ocean to the abnormal cooling was consisted in increasing of temperature and salinity under the surface in the Atlantic domain of the Nordic Seas due to Atlantic water sinking, as seen on the isopycnic surface 1028 kg/m^3 (Figure 5). Contrary, in the Arctic Domain (the Greenland and Iceland Seas) extremely cold and fresh surface waters were transported to greater depths. The isopycnic surface corresponds to the depth of 'overflow' waters. It means that the sharp decreasing of air temperature during the late 60^s caused both an intensification of convection in the Nordic Seas and changes of the overflow water properties. Indicated air temperature decrease seen both in the Mike and Torshavn meteorological station records (Figure 5). Positive subsurface temperature and salinity anomalies in the eastern areas of the Nordic Seas formed during this cooling event continued for almost 10 years and started to disappear with the emergence of water restratification triggered by the Great Salinity Anomaly propagation (GSA).

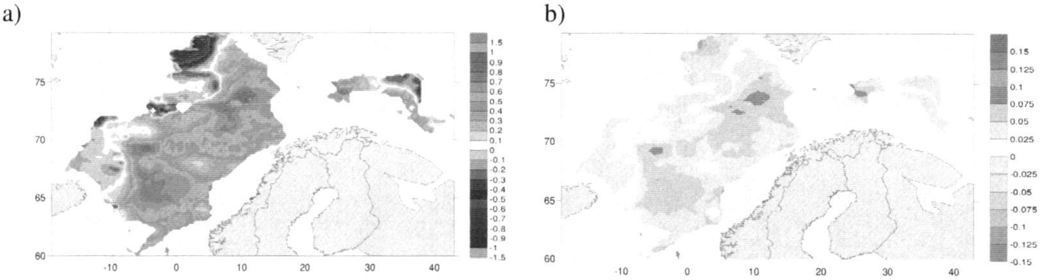

Figure 5. Temperature (a) and salinity (b) properties on 28.0 isopycnal surface (25-1000m)

The GSA entered the Nordic Seas at the end of 70^s and propagated to the Arctic for five years [2,3,4] since 1976 until 1981 (figure 4). New low salinity anomaly of Arctic origin was advected in the mid-90s. The late-90^s characterized by general temperature and salinity increase continued up to 2009.

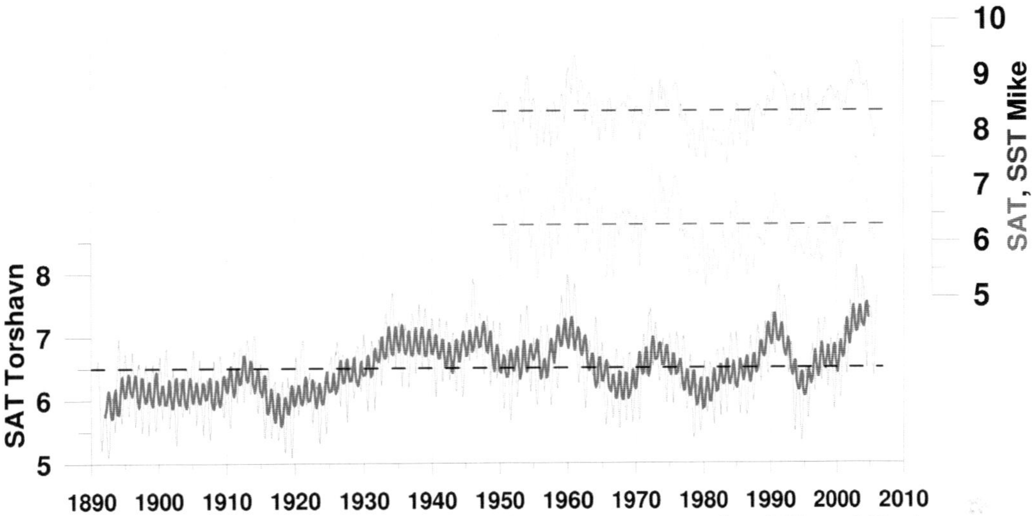

Figure 6. Sea surface temperature and sea air temperature on stations Mike and Torshavn

Low air temperature picks (figure 6) around the late-60^s, late-70^s and mid-90^s we regard as a consequence of the amplified ice export from the Arctic Ocean, what is confirmed by ice edge position migration [18,19]. Additional evidence comes from the Hadley Centre sea ice and sea surface temperature data set (HadISST) [16]. Each period of raised ice export was corresponded with the huge heat loss from the sea and MLD deepening.

6. Conclusions

The paper is focusing on the interrelationships between heat fluxes and the mixed layer depth. The heat fluxes and mixed layer depth were calculated by means of state-of-the-art algorithms. Historical oceanographic data for the Nordic Seas and OWS 'Mike' were extracted from ODB and combined with meteorological records. High correlation between heat fluxes and mixed layer depth has been found at OWS 'Mike' location for 2002-2009. Different hydrographic regimes in the Nordic Seas regulate the intensity of vertical exchange and properties of the upper mixing layer as seen from anomaly fields.

Temporal and spatial variability of main oceanographic parameters depict strong correlation between heat fluxes, the ML properties and large-scale characteristics of the water masses. The interrelation between air-sea heat fluxes and sea ice export from the Arctic Ocean has been confirmed. Both the mixed layer depth and the net heat flux are increasing during cold periods.

References

[1] Annear, R. L., and S. A. Wells (2007), A comparison of five models for estimating clear-sky solar radiation, Water Resour. Res., 43, W10415, doi:10.1029/2006WR005055

[2] Dickson, R. R., J. Meincke, S.-A. Malmberg, and A. J. Lee, 1988: The 'Great Salinity Anomaly' in the northern North Atlantic 1968–1982. Progress in Oceanography, Vol. 20, Pergamon, 103–151.

[3] Dooley, H. D., & Meincke, J. (1981). Circulation and water masses in the Faroese Channels during Overflow '73. Deutsche Hydrographische Zeitschrift, 34, 41–55.

[4] Dooley, H. D., Martin, J. H. A., & Ellett, D. J. (1984). Abnormal hydrographic conditions in the Northeast Atlantic during the 1970s. Rapports et Proce `s-Verbaux des Re ´unions du Conseil International polar Exploration de la Mer, 185, 179–187.

[5] Fairall, C. W., E. F. Bradley, D. P. Rogers, J. B. Edson, and G. S. Young, 1996a: Bulk parameterization of air–sea fluxes in TOGA COARE. J. Geophys. Res., 101, 3747–3767.

[6] Fairall, C. W., E. F. Bradley, J. E. Hare, A. A. Grachev, and J. B. Edson, 2003: Bulk parameterization of air–sea fluxes: Updates and verification for the COARE algorithm. J. Climate, 16, 571–591. Foltz, G. R., S. A. Grodsky, J. A. Carton, and M. J. McPhaden (2003), Seasonal mixed layer heat budget of the tropical Atlantic Ocean, J. Geophys. Res., 108(C5), 3146, doi:10.1029/2002JC001584.

[7] Hansen, B. (1985). The circulation of the northern part of the Northeast Atlantic. Rit Fiskideildar, 9, 110–126.

[8] Hansen, B. and Osterhus, S.: 2000, North Atlantic-Nordic Seas Exchanges, Progr. Oceanog., 45, 109-208;

[9] Hopkins, T. S. (1991). The GIN Sea — a synthesis of its physical oceanography and literature review 1972–1985. Earth-Science Reviews, 30, 175–318

[10] Johannesen, O. M. (1986). Brief overview of the physical oceanography. In B. G. Hurdle, The NordicSeas (pp. 103–127). New York: Springer.

[11] Josey, S. A., R. W. Pascal, P. K. Taylor, and M. J. Yelland.: 2003, A new formula for determining the atmospheric longwave flux at the ocean surface at mid-high latitudes, J. Geophys. Res., 108(C4), 3108, doi:10.1029/2002JC001418.

[12] Korablev A. A., Pnyushkov A.V., A.V. Smirnov (2007) Oceanographic database compiling for climate monitoring in the North European Basin. Trudy AARI, 447 (in Russian), pp. 85-107.

[13] Korablev A., A.Pnyushkov, A.Smirnov, O.Johannessen, V.Ionov (2006) Improved oceanographic database for the Nordic Seas. Geophysical Research Abstracts CD, Vol. 8, EGU General Assembly, ISSN: 1029-7006, EGU-A-04222.

[14] Papadakis, J. E., 1981: Determination of the wind mixed layer by an extension of Newton's method. Pacific Marine Sci. Rep. 81-9, Institute of Ocean Sciences, Sidney, BC, Canada, 32 pp.

[15] Price, J. F., R. A. Weller, and R. Pinkel, 1986: Duirnal cycling: Observations and models of the upper ocean response to diurnal heating, cooling, and wind mixing. J. Geophys. Res., 91, 8411-8427.

[16] Rayner, N. A.; Parker, D. E.; Horton, E. B.; Folland, C. K.; Alexander, L. V.; Rowell, D. P.; Kent, E. C.; Kaplan, A. (2003) Global analyses of sea surface temperature, sea ice, and night marine air temperature since the late nineteenth century J. Geophys. Res.Vol. 108, No. D14, 4407 10.1029/2002JD002670

[17] Reed, R. K.: 1977, On estimating insolation over the ocean, J. Phys. Oceanogr., 7, 482-485.

[18] Schmith, T., and C. Hansen (2003) Fram Strait ice export during the 19th and 20th centuries reconstructed from a multi-year dea-ice index from Southwestern Greenland, J. Climate, 16, 2782-2791.

[19] Spreen G., S. Kern, D. Stammer, R. Forsberg, and J. Haarpaintner (2006), Satellite-based estimates of sea ice volume flux through Fram Strait, Ann. Glaciol., 44, 321-328.

[20] Tanskanen, A., A. Arola, J. Kujanpaa (2003), Use of the moving time-window technique to determine surface albedo from the TOMS reflectivity data, In: Proc. SPIE Vol. 4896, p. 239-250.

[21] Thomson, R. E., and I. V. Fine (2003), Estimating mixed layer depth from oceanic profile data, J. Atmos. Oceanic Technol., 20(2), 319– 329.

Integral statistical eco-indices – effective complementary tool for assessment of ecological state of and ecological risks for water ecosystems

I. Bashamkova
Helsinki University, Helsinki, Finnland

Email: bashmako@mappi.helsinki.fi

Abstract. Eco-indices are successfully used for assessment of the ecological state and risks of water reservoirs. They allow, already at early stages, to detect negative effects on water ecosystems caused by progressive anthropogenic impacts and widening of the spectrum of pollutants, and to quantitatively evaluate ecological risks and damage for water reservoirs. Implementing these modern tools to water quality assessment is one of the lines to make decisions concerning challenging environmental problems.

1. Introduction

Human activities such as industry, agriculture and sewage water discharge have a considerable impact on water ecosystems and lead to their degradation. This impact causes suppressing the vital activity of water organisms and weakening natural self-purification that inevitably results in worsening the water quality with respect to all essential parameters of the water-quality standard. Good knowledge and quantitative evaluation of the environmental status of the ecosystem with regard to various types of pollutants is therefore a very important factor for the water management [9 – 11].

The existing governmental systems to assess the state of a water body suffer from a shortcoming that such assessments only reflect the possibilities of a water body use, but do not reflect its ecological condition. Therefore, efficiency of many water protection measures remains on a low level. Ecological assessment of anthropogenic impact on terrestrial surface waters includes estimation of adverse ecosystem changes on the basis of integrated analysis of both biotic and abiotic components. This method becomes more and more widespread in practice of the water quality assessment.

In view of progressive anthropogenic impact on water ecosystems the following first priority problems are relevant to management of complex natural systems:

1. Level of pollution;
2. Probabilities of extreme pollution events;
3. Risks for human health.

These problems are solved using modern statistical methods with the appropriate choice of the most sensitive indicator-organisms [8], [12].

2. Integral statistical eco-indices in assessment of ecological state of water ecosystems

At the present stage of the sanitary hydrobiology, the most prospective characteristics of the spatial and temporal distribution of ecological conditions in water reservoirs are integral indices.

They allow compression of *in situ* information, analysis of hydrobiological data with due account for hydrochemical and other data, which strongly increases the value of monitoring, assessing intensity of self-purification processes in water reservoirs, and quick and reliable determination of presence of toxicants in water.

With the proper expert selection of the representative parameters, eco-indices could have high sensitivities; and their use strongly facilitates and simplifies the diagnostics of pollutants in water [1], [13].

2.1 *Indices characterizing deviation of the local state of the water from its mean state*

Let us consider a water basin covered with M measurement stations $s = 1,2,...,M$. Each station s provides measurements of N parameters: $P_{sj}, j = 1,2,...,N$. The water-reservoir mean value P_j and the standard deviation σ_j of the parameter P_{sj} are calculated as

$$\overline{P_j} = \frac{1}{M}\sum_{s=1}^{s=M} P_{sj}, \quad \sigma_j^2 = \frac{1}{M-1}\sum_{s=1}^{s=M}\left(P_{sj} - \overline{P_j}\right)^2$$

We use σ_j as a natural scale for P_j and determine the dimensionless parameters:

$$D_{sj} = \frac{P_{sj}(s) - \overline{P_j}}{\sigma_j}.$$

The latter have mean values equal to zero and dispersions equal to unity. This allows us to sum them up into integral, multi-component indices (such as the bacterioplankton-development or the sanitary index) characterising each station of observations. For example, the eco-index E_s is determined as a linear combination of the dimensionless parameters D_i:

$$E_s = \sum_{j=1}^{j=N} \alpha_j D_j$$

Here, the coefficients α_j ($\sum \alpha_j = 1$) give weights to the parameters P_{sj}. Taking $\alpha_j = 0$ means nothing but neglecting the parameter P_{sj}. Determining the weight coefficients α_j (as well as the choice of parameters P_{sj}) is the subject of the hydrobiological expertise. If we have no grounds to distinguish between the more important and the less important parameters, the only reasonable idea is to prescribe equal weights to all parameters, that is to take $\alpha_j = N^{-1}$.

In our analysis we determine the basic parameters in such a way that their high values correspond to worse ecological situations and lower values - to a better situation [2–7].

2.2 *Indices characterizing deviation of the ecological state of a water reservoir from the officially established acceptable upper limit (AUL)*

Once again, we consider a water reservoir covered by M stations numbered as s = 1,2,...M, performing observations of N parameters P_{sj} (where j = 1,2,...N is the number of parameters, and s is the number of stations). The alternative dimensionless parameters are calculated as:

$$D_{sj} = P_{sj} / L_j$$

where L_i are acceptable upper limits (AULs) for P_j. Then we propose the following new index:

$$E_s^{(\lim)} = \sum_{j=1}^{j=N} D_{sj}\alpha_j$$

characterising the ecological state of the reservoir at each station. Here, the weight coefficients α_j ($\sum \alpha_j = 1$) are determined by an expert, accounting for the role of each parameter in the ecological health of the reservoir. The mean value of the index, namely,

$$\overline{E}^{(\lim)} = \frac{1}{M}\sum_{s=1}^{s=M} E_s^{(\lim)}$$

characterises the reservoir as a whole. By this means, the indices based on the AULs characterise the deviation of the factual ecological state of the reservoir from its still acceptable state. For details see [6] and [7]. In the traditional data analyses and interpretations aimed at assessments of the ecological state of water reservoirs [14, 15], these new methods (although quite simple and convenient) are still not applied.

3. Use of integral eco-indices for assessment of ecological risks for Early Warning of anthropogenic pollution of water ecosystems

Ecological risks are characterised by probabilities of situations leading to worsening of the state of natural ecosystems and/or their elements. The main objects, which suffer from negative anthropogenic and natural impacts, are natural ecosystems consisting of biotic and abiotic components. The very concept of "ecological risk" is relevant first of all to biota. The most challenging are risks of biodiversity degradation and deterioration of the environment dangerous for human health.

Water ecosystems are affected by the two main types of anthropogenic impacts: waste water discharges and transport of atmospheric pollutants through the water surface. Thus, analysis of the key mechanisms of water pollution and its ecological consequences, as well as assessments of ecological risks and damage for water bodies require interdisciplinary experience and joint efforts of experts in atmospheric sciences, hydrology and hydrobiology.

Eco-indices are successfully used for assessment of the ecological state of and ecological risks for the water reservoirs of different types. They allow, already at early stages, to detect negative effects on water ecosystems caused by progressive anthropogenic impacts and widening of the spectrum of pollutants, and to quantitatively evaluate ecological risks and damage for water reservoirs. Implementing these modern tools to water quality assessment is one of the lines to make decisions concerning the existing set of environmental problems and risk of accidents (droughts, floods and other natural disasters).

Eco-indices allow:

- To present basic results in a clear and illustrative way;
- To quickly and reliably detect deviations of water ecosystem from its acceptable state;
- To employ different systems of basic parameters – so that to address different tasks of monitoring;
- To essentially extend the list of indicators and to effectively use those that characterise functioning of hydrobionts missed in standard ecological monitoring;
- To establish hydrochemical parameters having the most significant influence on development of aquatic organisms;
- To compare inter-annual observations and to disclose basic trends and relationships in water ecosystems and water quality;
- To use integral indices for calibrating numerical models of water ecosystems

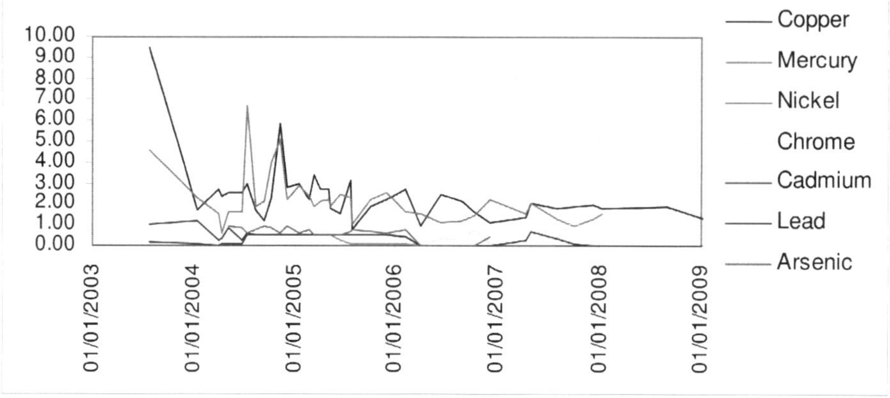

Figure 1. Temporal distribution of separate heavy metals (entire river Biferno, Italy).

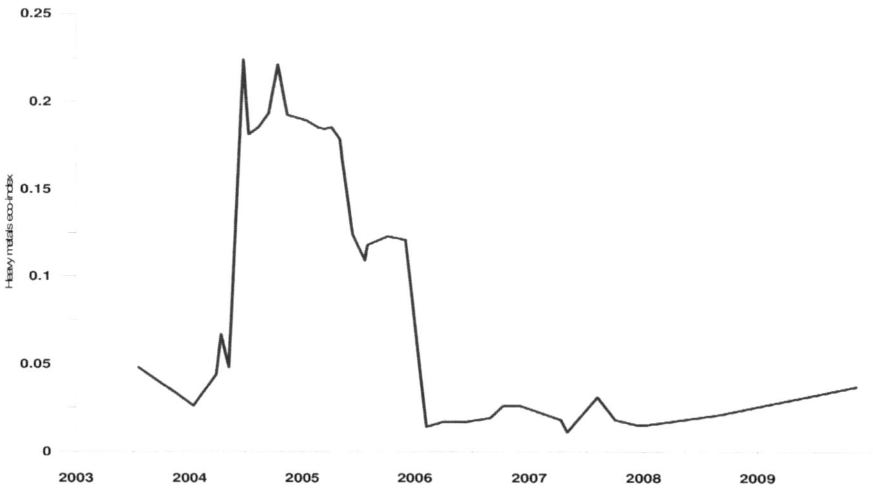

Figure 2. Temporal distribution of heavy metals index (entire river Biferno, Italy).

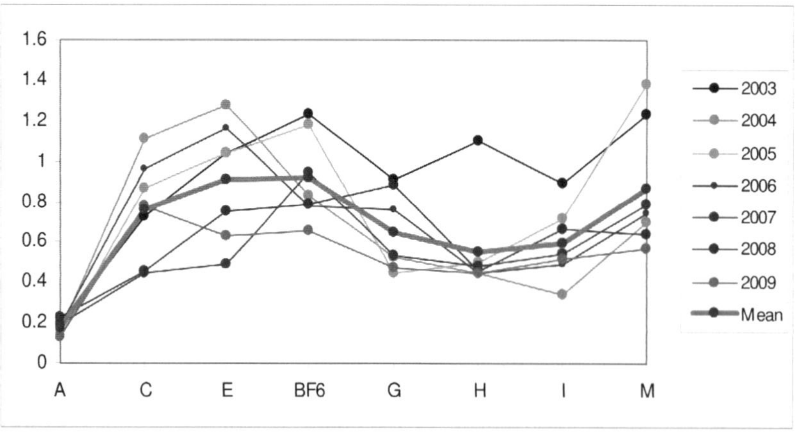

Figure 3. Long-term summer dynamics of the trophic eco-index for River Biferno (left to right - towards the see).

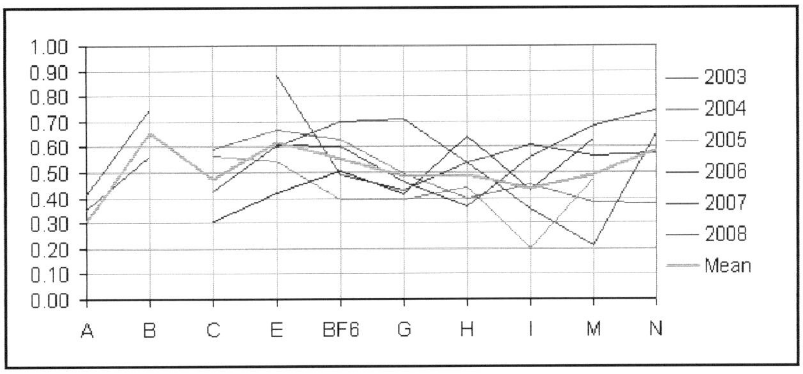

Figure 4. Long-term summer dynamics of the BOD5 eco-index for River Biferno (left to right - towards the see).

References

[1] Baumann M. and W. Popp, 1991. Bakteriologisch-hygienische Belastung ausgewahlter Oberflachengewasser und Klaranlagenablaufe in Bayern. *Berichte der Bayerischen Landesanstalt fur Wasserforschung* 18, Munchen.

[2] Bashmakova I., 2001. Modern state of biodiversity and the structure/functioning characteristic of bacterioplankton of the Danube Delta. *Ukrainian Hydro-ecological Society*, 4, 68-73.

[3] Bashmakova I., 2002. Eco-index for hydrobiological monitoring and water quality assessment with application to the Kilia Delta of River Danube. *34th International Conference IAD*, Tulcea, Romania, 27-30 August 2002, 463-469.

[4] Bashmakova I. and Levikov S., 2003. Statistical analysis and modelling of hydrochemical, hydrobiological and ecological conditions in the Danube River Basin. *35th International Conference IAD*, April 2003, Novi Sad, Yugoslavia.

[5] Bashmakova I. Kh., 2004. Ecological indices for estimation of ecological state and water quality in estuarine zones of big rivers. *Hydrobiol. J.*, 40, No. 3, 76-82

[6] Bashmakova I. and Levikov S., 2005. Application of integral indices to the assessment of ecological risks and damages. Extended Abstracts, *NATO Advanced Research Workshop on Air, Water and Soil Quality Modelling for Risk and Impact Assessment.* Tabakhmela, Georgia, 16-20 September.

[7] Bashmakova I. and Levikov S., 2007. Application of integral indices to the assessment of ecological risks and damages. In: "*Air, Water and Soil Quality Modelling for Risk and Impact Assessment*". Springer, 2007, 8.

[8] Baumert H., Levikov S., Pfeiffer K. D. and Goldman D., 2002. The MARS Data Collection and Analysis. Work Reports, HYDROMOD Scientific Consulting, Wedel, Germany.

[9] EU BATHING WATER QUALITY DIRECTIVE (1976): EU Bathing Water Quality Directive 76/160/EEC.

[10] EU BATHING WATER QUALITY DIRECTIVE (2006): EU Bathing Water Quality Directive 2006/7/EC.

[11] EU WATER FRAMEWORK DIRECTIVE (2000): EU-Water Framework Directive 2000/60/EC Council Directive 2000/60/EC of 23 October 2000: Establishing a framework for Community action in the field of water policy, 2000.

[12] Pollard, P., Huxham, M. 1998. The European Water Framework Directive: A new era in the management of aquatic ecosystem health? *Aquatic Conservation: Marine and Freshwater Ecosystems* 8, 773-792.

[13] Smith, V.H. 2003. Eutrophication of freshwater and coastal marine ecosystems: a global problem. *Environ Sci. Pollut.* R; 10: 126–39, 2003.

[14] Sokal, R.R. and F.J. Rohlf. 2000. *Biometry: The principles and practice of statistics in biological research*. W.H. Freeman and Co., N.Y., 887 pp.

[15] Van der Oost R., Vindimian E., van den Brink P.J., Satumalay K., Heida H., and Vermeulen N. P.E., 1997. Biomonitoring aquatic pollution with feral eel (Anguilla anguilla. III. Statistical analysis of relationships between contaminant exposure and biomarkers). *Aquatic Toxicology*, 39, 46-75.

IOP Conf. Series: Earth and Environmental Science **13** (2010) 012008 doi:10.1088/1755-1315/13/1/012008

Modelling the Impacts of Deforestation on Monsoon Rainfall in West Africa

B.J. Abiodun[1], J.S. Pal[2], E.A. Afiesimama[3], W.J. Gutowski[4], A. Adedoyin[5]

[1] Department of Environmental and Geographical Science, University of Cape Town, South Africa

[2] Department of Civil Engineering and Environmental Science, Loyola Marymount University, California, USA

[3] WMO Regional Research and Training Institute, Lagos, Nigeria

[4] Department of Geological and Atmospheric Sciences, Iowa State University, Iowa, USA

[5] Department of Physics, University of Botswana, Gaborone, Botswana

Email: babiodun@csag.uct.ac.za

Abstract. The study found that deforestation causes more monsoon moisture to be retained in the mid-troposphere, thereby reducing the northward transport of moisture needed for rainfall over West Africa. Hence, deforestation has dynamical impacts on the West African monsoon and rainfall.

1. Introduction

The West African monsoon circulation provides West African countries with more than 75% of their annual rainfall [1]. In summer, the monsoon accommodates important rainfall-producing systems: the African easterly jet (AEJ), tropical easterly jet (TEJ), African easterly waves (AEWs), and mesoscale convective systems (MCSs). These systems interact in a complex way with the low-level monsoon flow that transports moisture inland from the Atlantic Ocean to provide rainfall in West Africa. The dynamics of the monsoon and the associated rainfall pattern are sensitive to the changes in land use pattern [3]. However, the precise processes and mechanisms through which deforestation may impact the West African monsoon and rainfall are not yet fully understood; that is the focus of this study.

This study applied a dynamic regional climate model to investigate the impact of deforestation on monsoon and rainfall in West Africa. The climate model used is the International Centre for Theoretical Physics (ICTP) Regional Climate Model version 3 (RegCM3).

2. Simulations

For the study we analysed two RegCM3 simulations described in [2]. The model set-ups for the two simulations are the same, but the land use patterns differ, as shown in Figure 1. The first simulation, the control (CTR), used potential land cover (vegetation) conditions in West Africa. The second simulation (DEF), designed to study the effect of extreme deforestation over West Africa, use similar land cover with the control, but the forests are replaced with short grasses.

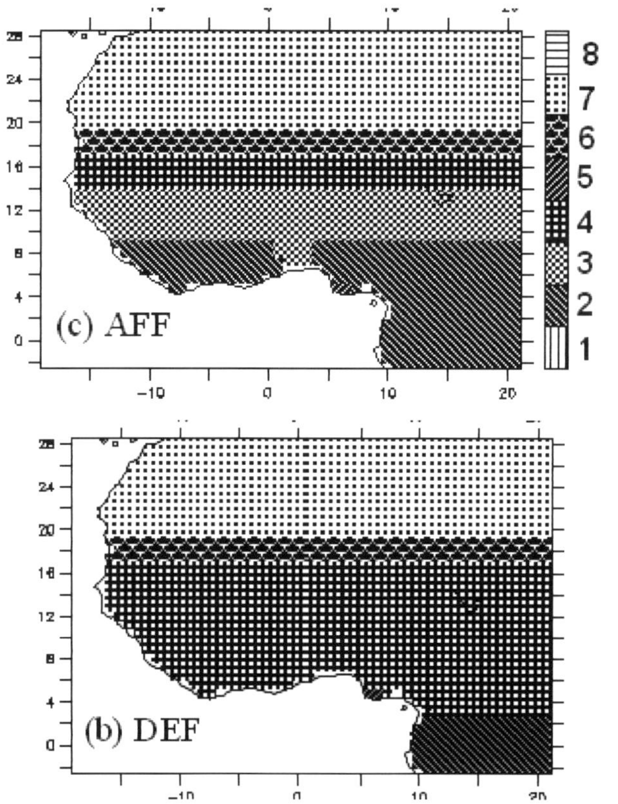

Figure 1. The distribution of the land cover type in West Africa used for the study: (a) Potential Vegetation experiments and (b) Deforestation (DEF). The land cover are: (1) disturbed forest, (2) evergreen broadleaf i.e. tropical forest, (3) tall grass i.e. woodland savannah, (4) short grass, (5) swamp, (6) semi-desert, (7) desert, and (8) crop/mixed farming.

3. Results and Discussions

The model captures all the essential climatic features in West Africa (Fig. 2 and 3). With the deforestation, rainfall decreases over the entire West African region (Fig. 2b). The distribution, however, generally remains the same when compared with the potential vegetation experiment. Fairly uniform reduction in rainfall (40% to 50%) is simulated from the coast to the Sahel.

The deforestation increases the monsoon flow south of 12° N (Fig 3a). This result differs from the deforestation simulation of [3], who obtain a decrease in monsoon flow due to deforestation. The simulated increase in monsoon flow here can be attributed to the reduction in surface roughness from deforestation [2]. Friction reduces the speed of low-level winds. In CTR, the friction from the forests decelerates the monsoon flow over the Guinean region, but in DEF, the deforestation reduces roughness length and displacement height by more than 90%. The reduction in surface drag experienced by the monsoon over the Guinean region in DEF is up to 0.04 (80%), causing the flow to move about 1.5 ms_1 (45%) faster than in the CTR simulation.

Consequently, the shallow circulation below 600 hPa is stronger and the mid-tropospheric easterlies, to the south of AEJ, increase by more than 1.5 ms^{-1} (Fig. 3c). In addition, the mid-tropospheric air in the region of increased easterlies is moister, but the boundary layer air is drier north of 9°N. This suggests that deforestation causes more monsoon moisture to be retained in the mid-troposphere, thereby reducing the northward transport of moisture needed for rainfall over West Africa. Hence, deforestation has dynamical impacts on the West African monsoon and rainfall.

IOP Conf. Series: Earth and Environmental Science **13** (2010) 012008 doi:10.1088/1755-1315/13/1/012008

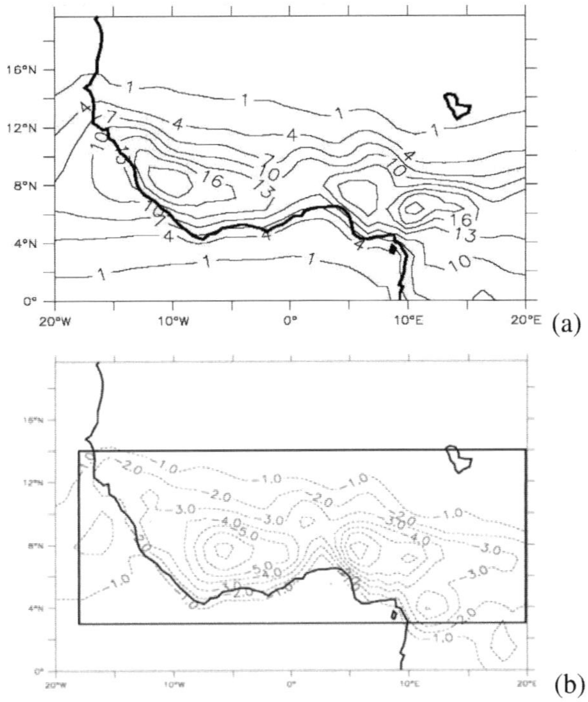

Figure 2. Simulate JAS rainfall in (a) CTR and (b) CTR - DEF.

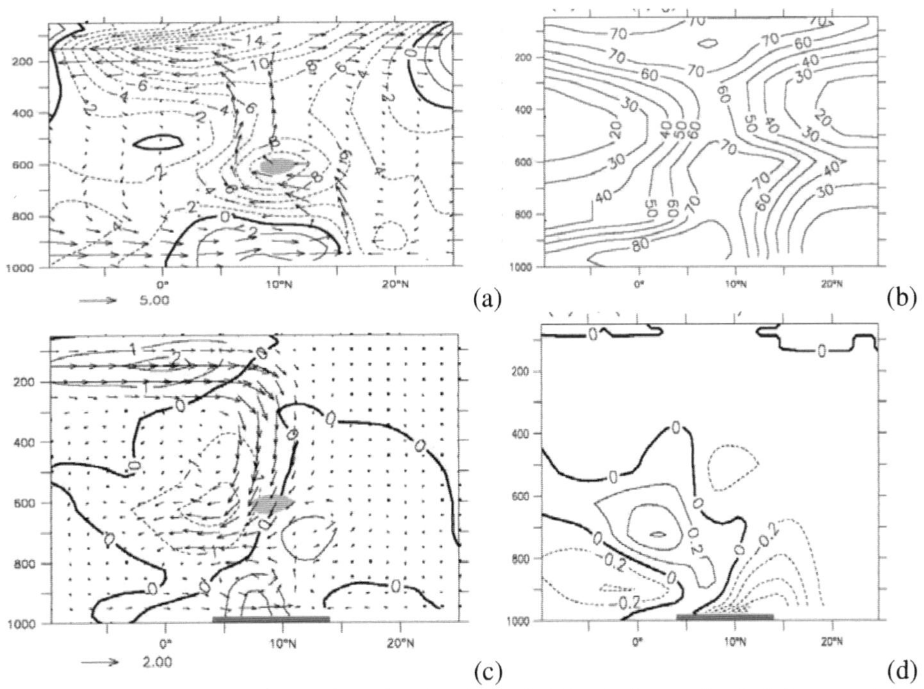

Figure 3. Simulated JAS vertical distribution of (a) wind: zonal wind (contour) with the core of AEJ (11ms_1) shaded, and meridional-vertical flow (vector); (b) relative humidity in CTR; differences in the simulated (c) wind and (d) specific humidity due deforestation (CTR-DEF) The shaded box in each panel indicates the location of deforestation area (i.e. approximately 4°–14°N).

References

[1] Omotosho J, 'Bayo 1985. The separate contributions of squall lines, thunderstorms and the monsoon to the total rainfall in Nigeria. *J Clim.*, **5**, 543–552

[2] Abiodun B.J., Pal J, E. A. Afiesimama, W. J. Gutowski and A. Adedoyin, 2008. Simulation of West African monsoon using the Regcm3 Part II: Impact of Deforestation and Desertification. *Theor. Appl. Clim.*, **93**, 245–261

[3] Zheng X.Y., Eltahir E.A.B. 1998. The role of vegetation in the dynamics of West African monsoons. *J Clim.*, **11**, 2078–2096

IOP Conf. Series: Earth and Environmental Science **13** (2010) 012009 doi:10.1088/1755-1315/13/1/012009

Impact of land cover changes on the South African climate

T.I. Ngwana[1], M.-E. Demory[2], P.L. Vidale[2], R.S. Plant[2], M.P. Mbedzi[3]

[1] *South African Weather Service, Pretoria, South Africa*
[2] *University of Reading, Earley Gate, Reading, UK*
[3] *Eskom Holdings, Cleveland, South Africa*

Email: isaac.ngwana@weathersa.co.za

Abstract. The Joint UK Land Environmental Simulator (JULES) was run offline to investigate the sensitivity of land surface type changes over South Africa. Sensitivity tests were made in idealised experiments where the actual land surface cover is replaced by a single homogeneous surface type. The vegetation surface types on which some of the experiments were made are static. Experimental tests were evaluated against the control. The model results show among others that the change of the surface cover results in changes of other variables such as soil moisture, albedo, net radiation and etc. These changes are also visible in the spin up process. The model shows different surfaces spinning up at different cycles. Because JULES is the land surface model of Unified Model, the results could be more physically meaningful if it is coupled to the Unified Model.

1. Introduction

The accuracy of numerical weather prediction relies on factors such as surface boundary conditions, for example, the prescription of wet initial soil moisture leads to considerable improvements to simulated precipitation (Beljaars et al. 1996). These conditions require specifications of the surface fluxes which account for all dynamics of the interactions between the land and the atmosphere. Previous studies (such as Anton, 1994 and Sun, 1994) have shown how the details of the fluxes can be modelled and which are meant to improve forecasts. The surface fluxes are influenced by the characteristics of the land (Lobocki, 2001). Therefore the impact of lower boundary conditions on modelling systems varies from region to region.

Various aspects of land surface sensitivity studies have been conducted. Sud et al. (1998) investigated the influence of land surface roughness on the large scale circulation. They found that the earth's vegetation cover has a large impact on the boundary layer water vapour transport convergence and the distribution of rainfall.

This influence of surface roughness on precipitation can be a localised process, for example, Sud and Smith (1985b) demonstrated that it affects summer rainfall over India. Other studies such as Wolfson et al, (1987) have demonstrated that soil moisture influences precipitation. Beljaars et al, (1996) revealed that precipitation of the of the wet initial moisture lead to the improvement in the simulated precipitation by the model. Over the Great Plains, for instance, its anomalous quantities enhance the reduction of rainfall while increasing the surface temperature. It is also important in the seasonal predictability. As shown by Fennessy and Shukla (1999), seasonal atmospheric pressure forecasts are improved by using realistic initial state of soil wetness. The above mentioned studies used global circulation models. In recent years however, such studies have been conducted using regional climate models.

South African Weather Service has more than two decade long numerical weather prediction history that began with the implementation and operational use of the ETA model (Riphagen et al,2002) and later the UK MetOffice Unified Model (UM). However, impact of fluxes on numerical weather prediction forecasts over the country has not been considered yet. South Africa has a strong seasonal cycle that greatly influences the characteristics of the land surface, particularly, vegetation. Therefore, surface variables, such as albedo, net radiation, soil moisture, and surface temperature are likely to be affected by this. In this we conduct a sensitivity study where we focus on the impact of these variables on the UM using the Joint UK Land Environmental Simulator (hereafter, JULES).

The structure of the paper is as follows: In the next section we discuss the models and the experiment design. The results are presented in section 3 and in section 4 we give our conclusions.

2. MODEL AND EXPERIMENT DESIGN

2.1 Model

In this study, we use the Joint UK Land Environmental Simulator (JULES 2.0). JULES is the offline version of the Met Office Surface Exchange System (MOSES) included in the Unified Model (UM) (Cox, 1999). It is also called "third generation" land surface model (Sellers, 1997). It represents land surface processes in natural ecosystems and includes the full hydrological cycle and vegetation effects on energy, water and carbon fluxes. It treats the land cover heterogeneity explicitly via nine land surface types (broad leaf, needle leaf, c3 grass, c4 grass, shrubs, inland water, urban, ice and bare soil). Its horizontal resolution is 1°x1° latitude and longitude and has four levels below the surface namely 0.10m, 0.25m, 0.65m and 2.0m. The land surface scheme can be coupled to Top-down Representation of Interactive Foliage and Foliage Including Dynamics (TRIFFID) model (Cox *et al.*, 1998), which updates both the plant distribution and the soil carbon based on climate-sensitive CO_2 fluxes at the land-atmosphere model interface and other vegetation processes (Cox, 2005). However, in order to simplify the analyses of the simulations in this study, the vegetation is kept constant. A full description of JULES can be found in Best (2005).

IOP Conf. Series: Earth and Environmental Science **13** (2010) 012009 doi:10.1088/1755-1315/13/1/012009

2.2 Study domain and land surface representation

Our domain of study is the South African region. This domain extends from 21.5˚S to 34.5˚S and from 17˚E to 34.5˚E (figure 1). This domain encompasses entire South Africa, Lesotho and Swaziland as well as some parts of Mozambique, Zimbabwe, Botswana and Namibia. However, only South Africa will be considered for the analysis. As we are only considering land points there are 213 grid points in the selected domain.

JULES surface representation is also shown in figure 1. It is clear from this figure that the model has very little or no needle leaf trees, inland water, urban and ice cover over South Africa. It only has broad leaf trees, C3 and C4 grass, shrubs and bare soil. The dominant surface type is tropical grass (C4) which covers most parts of the country. Other surface types which cover a significant part of the country are bare soil and shrubs (that cover most of the western part of the country). The eastern part of the country is covered mainly by C3 grass and broadleaf trees. There is generally a good agreement between the model vegetation cover and the actual vegetation cover over South Africa (see Figure 2). However, urban surface which dominates some parts of the country is not captured by the model.

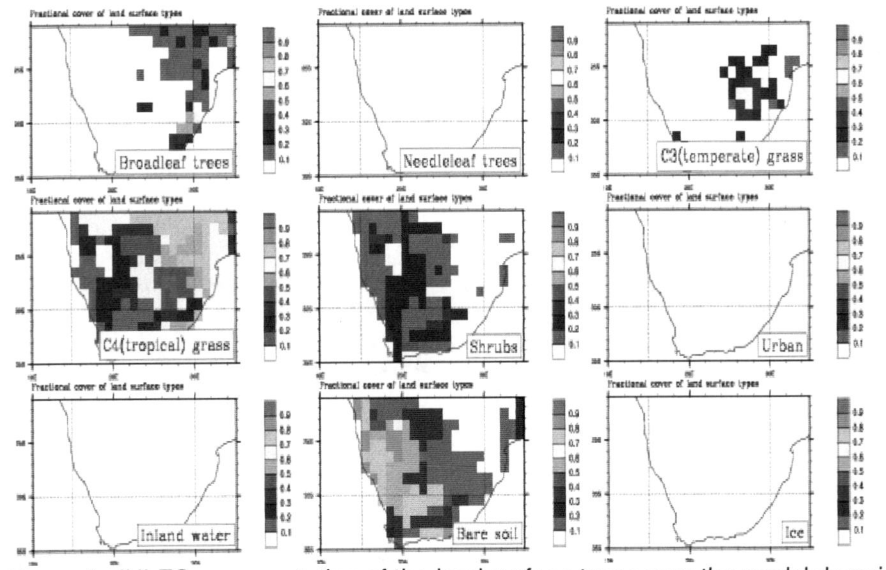

Figure 1: JULES representation of the land surface types over the model domain.

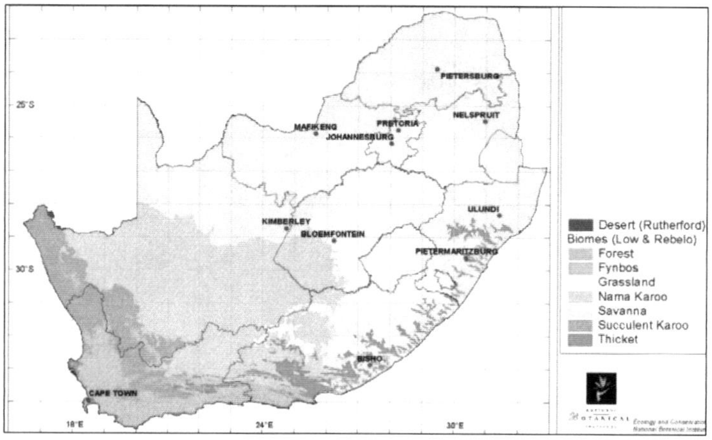

Figure 2: Vegetation cover over South Africa (Low and Rebelo, 1996)

2.3 Experiments

The simulations were performed over the domain of study. The atmospheric data used to drive JULES have been obtained from the Global Soil Wetness Project (GSWP-2). These include 2-metre temperature, 2-metre specific humidity, 10-metre winds, shortwave down and longwave down at the surface, surface pressure, rainfall rate (convective and total) and snowfall rate.

Vegetation was set to be static (without phenology). In all experiments, actual land surface cover in the model was replaced by one single land surface type over the entire domain. We performed 5 experiments: 1) the entire domain was covered with bare soil only, 2) tropical grass (C4) only, 3) shrubs only, 4) broadleaf trees only and 5) a control experiment. Shrubs and tropical grass were chosen for this investigation because they cover a large part of South Africa. Broadleaf trees were chosen because they make the greatest contrast with bare soil which is also cover significant part of the South Africa. Ice, needle leaf trees, inland waters and urban surface types were not considered in the present study because they either cover a small part of South Africa or are not present at all.

2.4 Model spin up and its results

To bring JULES into equilibrium after changing the surface types and in control run, a spin up was run. It uses temperature and soil moisture as convergence criteria. Convergence criteria for soil temperature were set to 0.1 Kelvin and 1% for soil moisture. For spin up to be considered successful, soil moisture and soil temperature change between the cycles should not exceed the criteria set. But where there are very few points not spun up and where the model shows stability, it can also be regarded as spun up. For example, in the shrub surface (where soil moisture is a criterion) the model becomes stable after eleven cycles with only one point failing to meet convergence criterion and in control surface it becomes stable after twenty one cycles with two points not converged out of 213 points.

Spin up was run for several cycles until the set criteria are met. After the model reaches equilibrium, it continues with normal run. That is, the spin up phase is set to precede the main run. Each experiment was carried out with its own separate spin up. JULES spin up results for this study are shown in Figures (3a & b).

It can be seen from these figures that the spin up period and spin up sequence for various surface types are different. The model spins up quicker when using soil temperature criterion than when soil moisture is used as a criterion. For example, in Figures (3a & b) the model span up faster in broadleaf surface and slower in shrubs and control surfaces. Spin up for bare soil surface also slower than the one for broad leaf surface.

Spin up results indicates how rapidly the land-atmosphere responds to changes in land use. If we consider soil moisture convergence, we find that vegetation surface promotes interactions between the atmosphere and the surface by extracting moisture from the soil and sending it to the atmosphere. For example, broadleaf trees have the maximum root depth of about three metres; hence they can extract soil moisture from the depth of three metres below the surface. This explains reasons for broadleaf trees surface spins up faster than any other surface type. We can see again that shrubs bring less moisture out of the soil than C4 grass because shrubs have higher stomata resistance than C4 grass. The less stomata resistance in c4 grass promotes the transfer of moisture between soil and atmosphere as compared to shrubs which has higher stomata resistance. This leads model to spin up faster in C4 grass surface than in shrubs surface. In general the model spins up twice or more slower in the bare soil surface compared to broad leaf surface because bare soil has no extra means that can aid to extract moisture from the soil into the atmosphere other than evaporation.

In contrast to soil moisture heat enters and leaves the surface easier and faster. Heat is easily transferred between soil layers as well as between the surface and the atmosphere. Hence the model spins up quicker in soil temperature than in soil moisture. From the spin up results we can conclude that surface type plays a significant role in heat and moisture balance between the surface and the atmosphere.

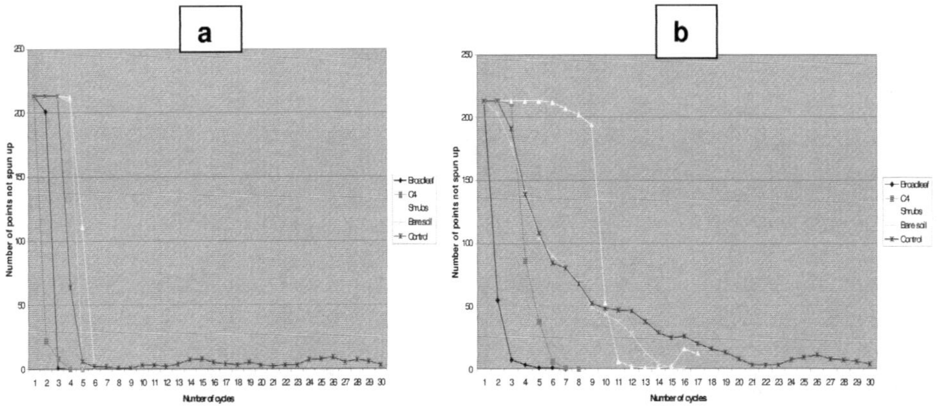

Figure 3: JULES spin up cycles based on (a) soil temperature and (b) soil moisture for control, C4 grass, broadleaf trees, shrubs and bare soil.

IOP Conf. Series: Earth and Environmental Science **13** (2010) 012009 doi:10.1088/1755-1315/13/1/012009

3. Model simulations results

JULES simulations of surface temperature over South Africa are very close or similar to the observed (Figure 5(a) and (b)). These results give us confidence that JULES performs well over South Africa. As a result, difference between control and experiments over South African domain can be useful and physically meaningful.

The model results have been averaged for 13 years (from 1982-1995). Simulation from each surface type is compared with the control run. From Table 1 we can see how experiment results deviates from the control results. On average, albedo has decreased in all surface types with largest decrease of about 31% in broadleaf surface. The most significant net radiation change is seen from the broadleaf surface which may have resulted in the highest temperature change (1K). The change in both runoffs and subsurface runoff is so small that we decided not to include them in the table.

Total soil moisture has increased in all surface types with the highest increase in bare soil (increase of about 107%) followed by shrubs surface (about 5 %). Soil moisture change in C4 grass and broad leaf surface is very small. The larger increase of soil moisture in bare soil surfaces have been caused by the fact that apart from evaporation which mainly takes at the top layer and subsurface runoff, there is no other mechanism that extracts moisture from the soil. The increase in soil moisture over shrubs surface may have been caused by the fact that shrubs restrict evapotranspiration by its strong stomata resistance and roots cannot extract soil moisture from layer third and the fourth soil surface layers since its root depth is 0.5m.

Table 1: JULES simulations averaged over SA for 1983 to 1995 period

Surface Type	**Albedo**	**Net Radiation (Wm^{-2})**	**Temperature (K)**	**Total soil moisture**
Control	1.05	52.38	293.5	290.51
Broadleaf Trees	0.73	58.02	294.5	290.71
C4 grass	1.02	53.16	293.6	291.89
Shrubs	0.93	52.38	293.5	306.47
Bare soil	1.02	52.14	293.4	600.6

IOP Conf. Series: Earth and Environmental Science **13** (2010) 012009 doi:10.1088/1755-1315/13/1/012009

Seasons of the year have shown to be playing a role in controlling surface temperatures (figure 4d). The albedo (which is not shown here) for each surface does not change with time. This is meaningful since we used static surface types.

If we are considering simple general relationship amongst albedo, net radiation and surface temperature, we expect high albedo to bring about low net radiation which inturn results in low temperature and low albedo to increase net radiation which leads to high temperature. This is shown for example in broadleaf surface which has lowest albedo, highest net radiation and highest surface temperature. This simple explanation is not applicable to bare soil and other surface types. This suggests that JULES calculates surface temperature in a complex way that it cannot simply explained from net radiation point of view alone. For instance, one of the factors which increases surface temperature over a vegetated surface is that leaves of trees can prevent some of the infrared radiation from escaping the surface.
The fact that broad leaf has highest latent heat throughout the whole year (Figure 4a) agrees well with the fact that it has the lowest soil moisture content (Figure 4f) throughout the whole year and bare soil which has the lowest latent heat has the highest soil moisture content. Seasonal change of surface runoff (Figure 4f)) is more prominent on bare soil surface type than any other surface type. Surface runoff is highest on bare soil surface. It peaks during rainfall season (summer) and drops in every surface type during winter. This shows that the model considers surface runoff as an immediate respond to the rainfall and on the other hand soil temperature does not seem to respond immediately. The amount of water in the soil that cannot evaporate or go to subsurface runoff is stored in the soil and forms part of total soil moisture. This gives us the reason why there is highest of total soil moisture (Figure 4f) over the bare soil surface through the whole year.

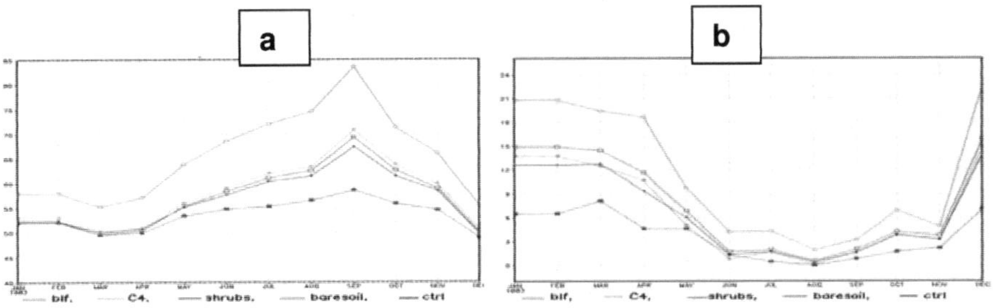

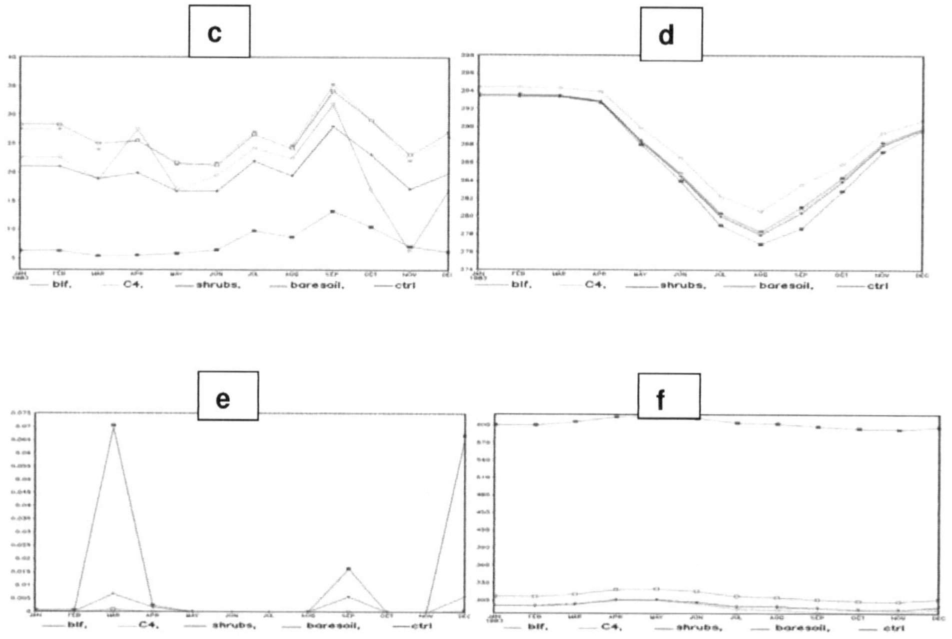

Figure 4: seasonal cycles of net radiation(W/m^2)(a), latent heat (W/m^2) (b), sensible heat (W/m^2) (c), surface temperature (K) (d), Surface runoff(g/m^3) (e), and, total soil moisture (g/m^3) (f)

Surface albedo of broad leaf surface has decreased over the whole domain. Net surface radiation for broad leaf has increased everywhere (Figure 6a). C4 grass and shrubs show net surface radiation increase in the west and decrease in the east (Figure 6b-c). Bare soil also displays small patch increase in east and decrease everywhere else (Figure 6d). This explains in simple terms the general effect of surface albedo in net surface radiation.

Surface temperature demonstrates its direct relationship with net radiation. Where there is net radiation increase there is positive surface temperature anomalies. For example, in Figure 7a we can see that broad leaf surface type has positive surface temperature anomaly everywhere which correspond with the increase in net radiation that is shown in figure 6a. This relationship is also seen on bare soil and more conspicuous on c4 and shrubs surfaces where we find that there are positive temperature anomalies (Figure 6a-d) in areas where there are positive net radiation (Figure 7a-c).

Soil moisture in bare soil (Figure 8d) and shrubs surfaces (Figure 9c) increase as we move towards the east, C4 total moisture (Figure 8b) increased in the central part and broad leaf has negative anomaly almost everywhere. Negative total soil moisture anomalies over broadleaf surface can be ascribed to the fact that broadleaf trees have the ability to remove the soil water from deep soil and release it to the atmosphere as stated above. For the same reasons stated above bare soil surface has high moisture content (Figure 8d) and less moisture content in broad leaf surface (Figure 9a).

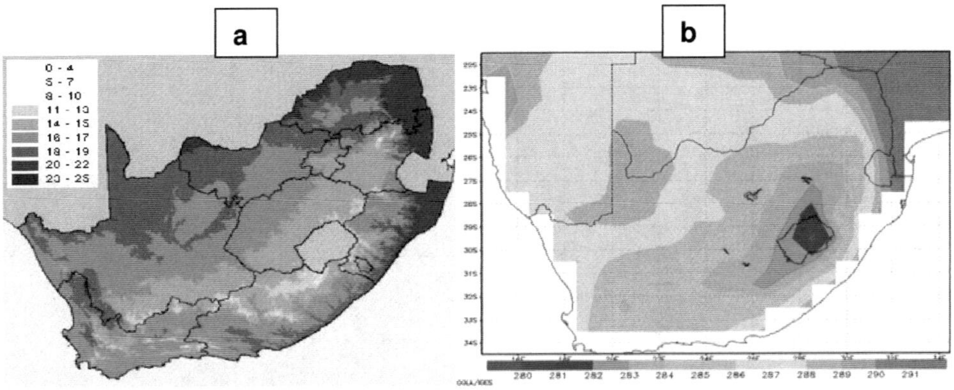

Figure 5: Observed surface temperature (a) in degrees Celcius and surface temperature simulated by JULES (b) in Kelvin

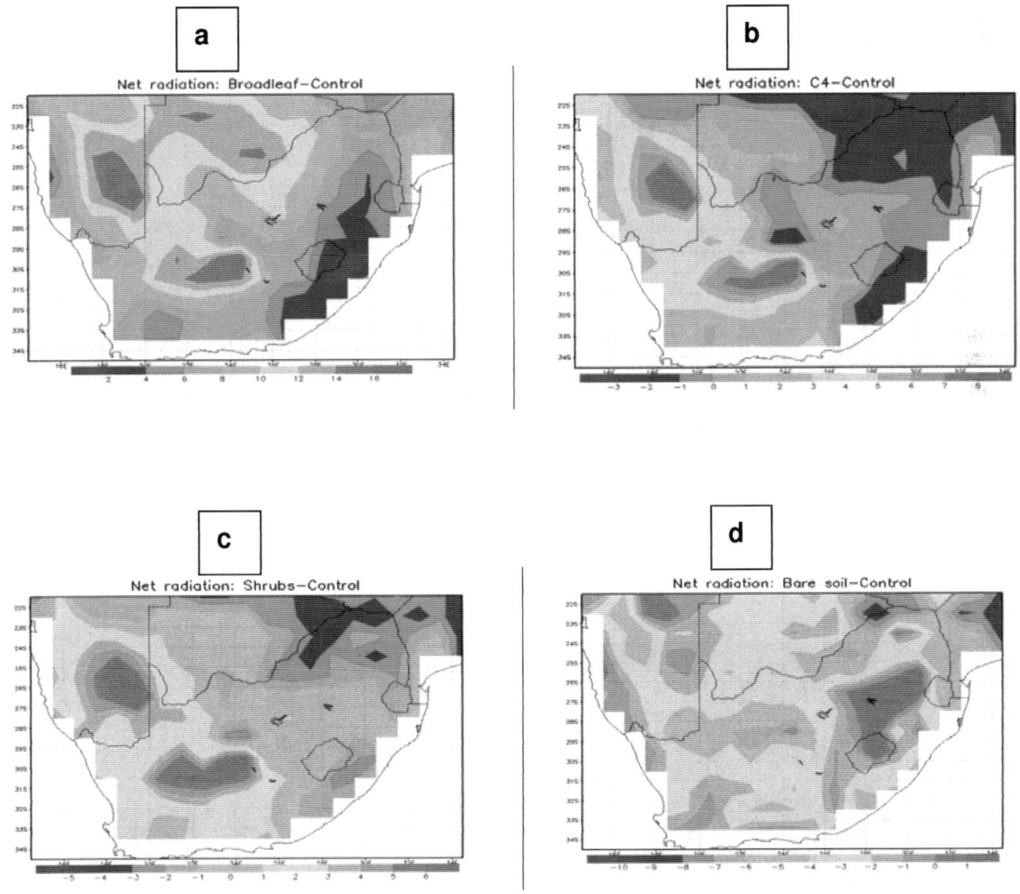

Figure 6: Net radiation (W/m^2) anomalies for broadleaf (a), c4 grass (b), shrubs (c) and bare soil (d)

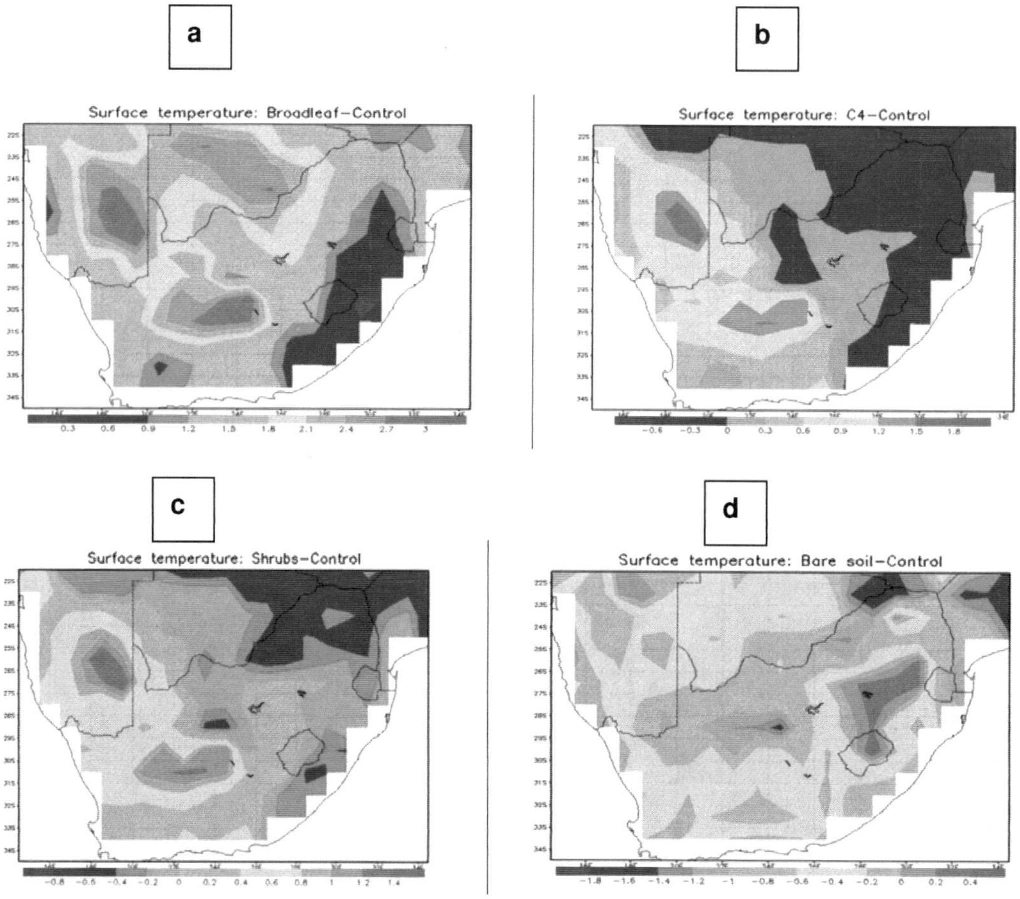

Figure: 7 Surface temperature anomalies for broadleaf (a), c4 grass (b), shrubs (c) and bare soil (d)

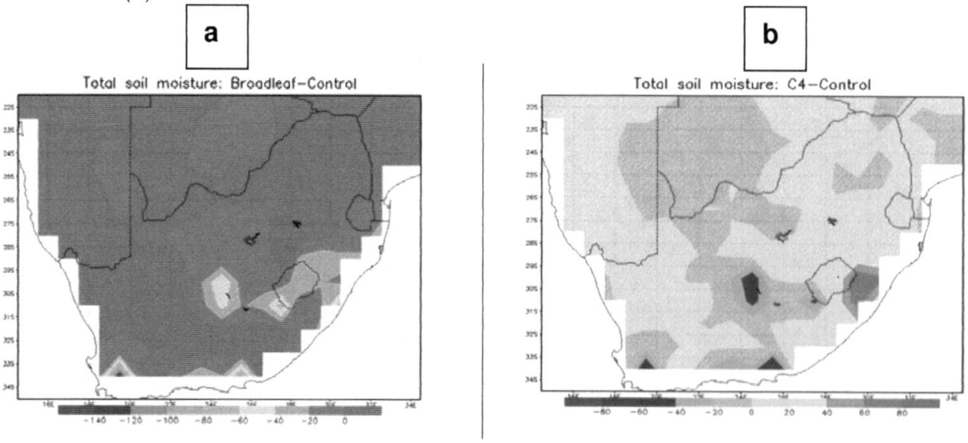

IOP Conf. Series: Earth and Environmental Science **13** (2010) 012009 doi:10.1088/1755-1315/13/1/012009

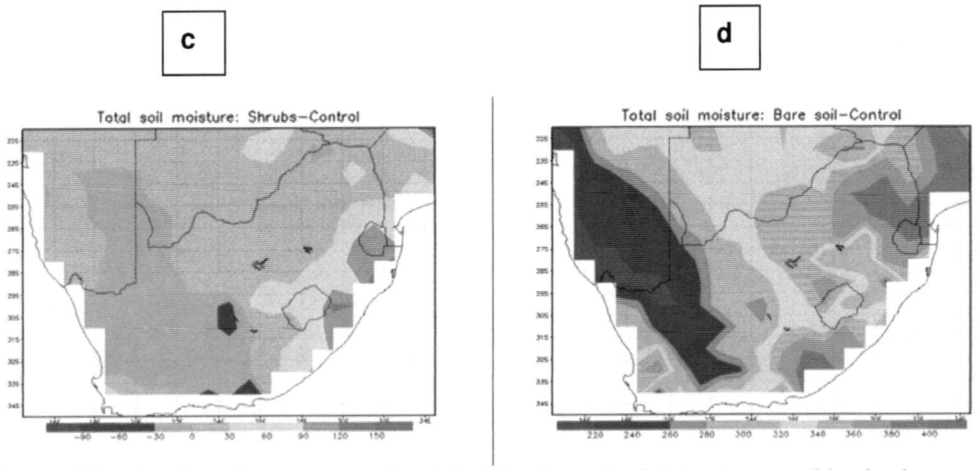

Figure:8 Total soil moisture anomalies (g/m3)for broadleaf (a), c4 grass (b), shrubs (c) and bare soil (d)

4. **Conclusions**

The main aim of this project was to investigate if JULES is sensitive to the change in the land surface types in South Africa. Then experiments which involved the changing of land surface cover of the model into one homogeneous type were conducted. The model was allowed spun up first before the actual run is made. The model spin up results show that JULES is sensitive to land surface change. This is noticed from the fact that the model span up at different cycles for each surface type. Broadleaf surface type was seen to be the surface which highly promotes the interaction between land and atmosphere than any other surface. This was evident when the model spun faster in broad leaf surface type than any other surface.

Geographic location of an area plays a role in factors such as surface temperature, moisture and etc. This is revealed when homogeneous surface brings about different amount of change of variable in the domain. For example, we have notice that one homogeneous surface bring positive anomalies in one area and negative in the other.

The model again demonstrated that changes that are brought about by changing the land surface types are also influenced by seasonal cycle. These model results suggest that if there can be a severe desertification, it can have a serious effect on the energy balance and soil moisture over the country. For example there can be flood increase when the surface is change to bare soil since soil moisture in deep layers will not be brought back to atmosphere and flooding will increase significantly especially during summer (rainfall) season due to the increase in surface runoff. However if the vegetation cover can be extended to large area, it will reduce flooding and will increase moisture in the atmosphere through evapotranspiration which may lead to the increase of precipitation.

The model also managed to show the great contrast between broadleaf surface and bare soil surface. The model has shown to be more physically consistent regarding albedo, net radiation and surface temperature.

We envisaged that the next phase of the project will coupled JULES to the Unified Model to determine the extent in which surface change can have on large scale features, rainfall, surface pressure and etc.

Acknowledgements. The authors would like to express their appreciation to Dr Warren Tennant who came with the idea of the project as well as to the UK Meteorological Office and the South African Weather Service for sponsoring this project.

REFERENCES

Best M, 2005: JULES Technical Documentation. UK Met Office, Fitzroy, Exeter, UK

Beljaars A.C.M.,1994: Parameterisation of Surface fluxes in large-scale models under free convection, *Quart.J.Royal.Meterol.Soc.**121**, 255-270*

Charney, J.G., 1975: Dynamics of desert and drought in the Sahel. *Quart.J.Royal.Meteor.Soc.* ,**101**, 193-202.

Cox, P.M., R.A. Betts, C.B. Bunton, R. L. H. Essery, P.R. Rowntree, and J. Smith, 1999: The impact of new land surface physics on the GCM simulation of climate and climate sensitivity. *Climate Dynamics.*,**15**, 183-203

Essery, R.L.H., M.J. Best,, R.A. Betts., Cox, P.M. and Taylor, C.M., 2003: Explicit Representation of Subgrid Heterogeneity in a GCM Land-Surface Scheme, *J Hydrometer.*, **5**, 30-543

Lobocki, L.M, 2001: Calculation of Surface Fluxes under Convective Conditions by Turbulence Closure Models. *Amer Met Soci., Vol* **40.**,604-621

Low, S.B and A.G Rebelo., 1996: Vegetation of South Africa, Lesotho and Swaziland, Department of Environmental Affairs and Tourism, Pretoria.

Riphagen H.A, Bruyere, C.L, W.Jordaan, Poolman, E.R and Gertenbach, J.D: 2002. Experiments with the NCEP Regional Eta Model at the South Africa Weather Bereau, with Emphasis on Terrain Representation and Its Effect on Precipitation Predictions, *Mon Wea Rev,* Vol 130, 1246-1263

Shaw, W.J and Doran, 2001: Observations of systematic boundary layer divergence patterns and thear relationship to land use and topography. *J. Climate*, **14**, 1753-1764

Sun J, and Mahrt L.,1994: Determination of surface Fluxes from the surface Radiative Temperature. *J. Atmos. Sci.* Vol **52**, No 8

Van Heerden, J., D.E. Terblanche, and G.C. Shulze, 1988: The Southern Oscillation and South African summer rainfall. *J.Climatol.*, **8**, 577-597.

Wei, H., and C. Fu., 1998: Study of the sensitivity of a regional model in response to land cover change over northern China. *Hydrol. Process.*, **12**, 2249-2265.

IOP Conf. Series: Earth and Environmental Science **13** (2010) 012010 doi:10.1088/1755-1315/13/1/012010

Determining astronomical seeing conditions at Matjiesfontein by optical and turbulence methods

M Nickola[1], I Esau[2] and G Djolov[3]

[1] Hartebeesthoek Radio Astronomy Observatory (HartRAO) Space Geodesy Programme, PO Box 443, Krugersdorp 1740, South Africa
[2] G.C. Rieber Climate Institute of the Nansen Environmental and Remote Sensing Center (NERSC), Thørmohlensgate 47, N-5006 Bergen, Norway
[3] University of Pretoria, Department of Geography, Geoinformatics and Meteorology, Faculty of Natural and Agricultural Sciences, University of Pretoria, Pretoria 0002, South Africa

Abstract. Matjiesfontein in the Karoo has been proposed as a suitable location for a new fundamental space geodetic observatory. On-site geodetic equipment will include a Lunar Laser Ranger (LLR). LLR requires sub-arcsecond optical seeing conditions for delivery of high quality and quantity data. Seeing conditions at the Matjiesfontein site will be evaluated by making use of an automated seeing monitor and by modelling atmospheric turbulence with Large Eddy Simulation Nansen Center Improved Code (LESNIC).

Introduction

The Space Geodesy Programme operates from the HartRAO site where geodetic data quality and quantity are adversely affected by pollution, RFI (Radio Frequency Interference), cloud cover and obsolete instrumentation. A new fundamental space geodetic observatory at a new site with new equipment has been proposed. A geotechnical survey has identified a site just 4 km south of Matjiesfontein in the Karoo as suitable for optimal scientific output.

An S/LLR (Satellite / Lunar Laser Ranging) system is to be a major addition to the equipment to be located at the new space geodetic observatory. Because of single-photon returns, LLR requires optimal optical seeing conditions, which will allow for the propagation of a laser beam through the atmosphere without excessive beam degradation. A site suitable for LLR must deliver seeing of ~1 arc-sec for the LLR to achieve millimetre-level accuracy in ranging to the Moon. [1]

Figure 1. Matjiesfontein site – looking north towards proposed LLR location on ridge.

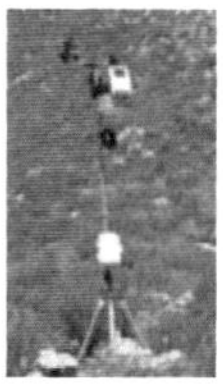

Figure 2. On-site Vantage Pro 2 weather station.

Figure 3. Matjiesfontein site – looking southeast from LLR ridge down into valley.

IOP Conf. Series: Earth and Environmental Science **13** (2010) 012010 doi:10.1088/1755-1315/13/1/012010

Seeing

A star is located at an immense distance from the Earth and should appear as a point source when observed through a telescope (or by the naked eye). However, it is observed as a blurred, moving image. The main contributor to the distortion of the stellar image is atmospheric turbulence.2]-[7]. Turbulent layers in the atmosphere cause index of refraction variations that distort the plane wave fronts that ideally should be observed.[4]-[7] A telescope's theoretical angular resolution may be smaller than an arc-sec, but the image resolution will be limited further by atmospheric seeing conditions. Seeing may be quantified: (2.1) theoretically; (2.2) by measurement; (2.3) by combining techniques –

1.1. Theory: modelling with LESNIC

The atmospheric boundary layer is described by a set of differential equations and appropriate boundary conditions. Local unstable air masses (eddies) are either resolved (large eddies) or modelled (small eddies).8]. Assuming a Kolmogorov model of turbulence [9], a profile of turbulence strength as a function of altitude, $C_{\mathrm{N}}^2(h)$, may be determined –

$$C_{\mathrm{N}}^2(h) = \left(80\times10^{-6}\frac{P(h)}{T^2(h)}\right)^2 \; C_{\mathrm{T}}^2(h) \tag{1}$$

where $C_{\mathrm{T}}^2(h)$ is the temperature profile of the atmosphere, h the height above the telescope, P is the pressure in hPa and T the absolute temperature in K [10]. The C_T^2 parameter is obtained from the rates of temperature variance dissipation, ε_θ and turbulence kinetic energy dissipation, $\varepsilon^{-1/3}$,

$$C_T^2 = 1.6 \;\; \varepsilon_\theta \;\; \varepsilon^{-1/3} \tag{2}$$

The LESNIC model will be employed to deliver the profile of turbulence strength as a function of altitude (also referred to as the index of refraction structure parameter or structure constant).

1.2. Preliminary results using LESNIC

The test for the large-eddy simulation technique for determining seeing conditions has used the DATABASE64, which consists of a collection of LESNIC runs for a stably stratified planetary boundary layer (SBL) over a homogeneous aerodynamically rough surface [16].

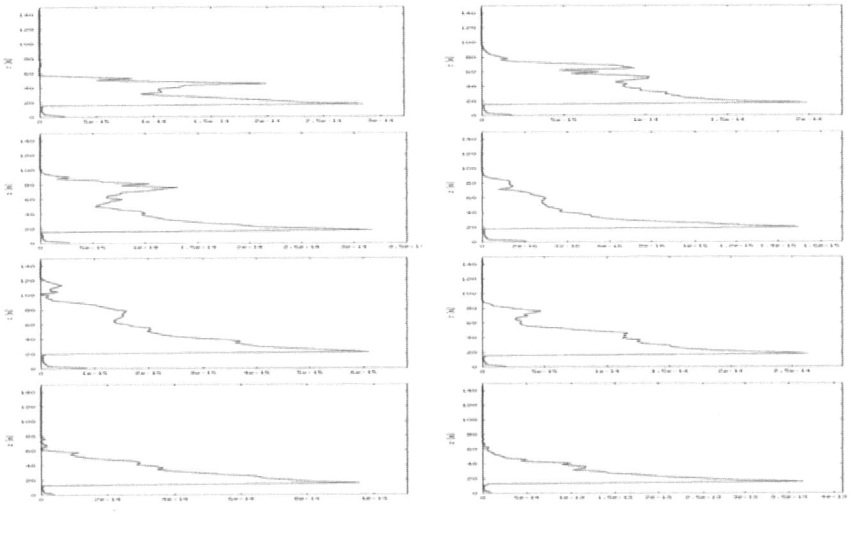

Figure 4. Profiles for $C_N^{\,2}(h)$ obtained with LENIC.

Figure 4 presents eight runs with LESNIC, which cover range of stability conditions in PBL. The quantitative shape of the $C_N^{\,2}(h)$ profiles is quite reasonable.

In Figure 5 the comparison of one of the LESNIC runs with data from [17]. This paper presents 3 1/2 years of site testing data obtained at Dome C, Antarctica, based on measurements obtained with three DIMMs located at three different elevations. Basic statistics of the seeing and the isoplanatic angle are given, as well as the characteristic time.

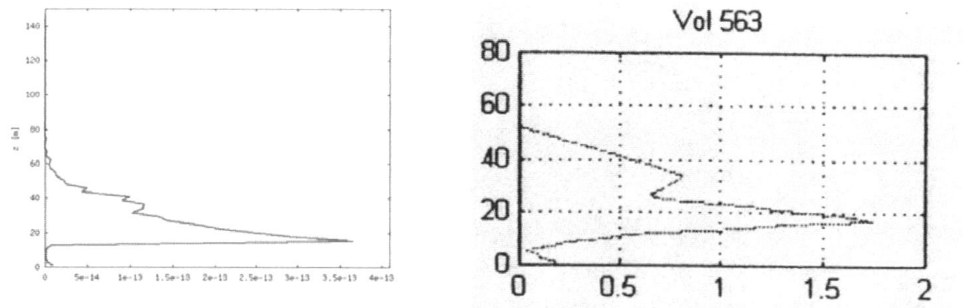

Figure 5. $C_N^{\,2}(h)$ profiles (left panel obtained with LESNIC), right profile from [17] (The horizontal axis is in units of 10^{-13} $m^{-2/3}$).

The figure illustrates LESNIC capacity of reproducing observed features of seeing conditions. The future run with LESNIC will use the representative to Matjiesfontein site initial and boundary conditions as well as topography and roughness features. Therefore, the use of the large-eddy simulation technique will allow conducting relevant numerical simulations for determining the long term seeing conditions at the proposed site as well as the conditions for a particular observation periods.

1.3 Measurement: seeing monitor

By determining the smallest resolvable angle, atmospheric seeing conditions can be measured [7]. The system consists of a telescope, CCD camera and control/processing PC. Seeing varies spatially as well as temporally.[6],[7] Stars at various positions on the night sky will be observed from various on-site locations. Seeing varies at a rate of more than 100 times a second. Numerous short exposure images of a specific source will be captured, evaluated individually, as well as stacked to obtain an average output image, the Point Spread Function (PSF). The PSF is broadened by poor seeing conditions.[4] It will be compared with the theoretical Airy function, which describes the image under ideal conditions (i.e. how the image would appear outside the Earth's atmosphere). A quantitative measure of the seeing will be obtained by fitting a Gaussian intensity distribution to the PSF and determining the Full Width at Half Maximum (FWHM). The FWHM of a star's intensity distribution at the focus of the telescope describes the seeing. It is given by the standard deviation from the Gaussian distribution, σ [11] –

$$\varepsilon_{\mathrm{FWHM}} = \mathrm{FWHM} = 2\sqrt{2\ \ln 2}\ \sigma \approx 2.355\ \sigma \qquad (3)$$

The Fried parameter, r_0,[12],[13] is a statistical parameter to characterise the seeing and is related to the C_N^2 profile as follows –

$$r_0 = \left(16.7\lambda^{-2} \frac{1}{\cos\gamma} \int_0^\infty C_N^2(h)\ \mathrm{d}h \right)^{-3/5} \tag{4}$$

where the turbulence strength $C_N^2(h)$ varies as a function of height h above the telescope, γ is the zenith angle and λ is the wavelength of observation.[2] A larger Fried parameter indicates better seeing conditions.

Seeing is related to the Fried parameter, r_0, by –

$$\varepsilon_{\mathrm{FWHM}} = 0.98\ \frac{\lambda}{r_0} \tag{5}$$

where λ is the wavelength of observation.[12],[13]

From Equations (3) and (4) above, the relation between seeing and the profile of turbulence strength as a function of altitude, $C_N^2(h)$, follows [14]–

$$\varepsilon_{\mathrm{FWHM}} = 0.98\ \frac{\lambda}{r_0} = 5.25\lambda^{-1/5} \left(\int_0^\infty C_N^2(h)\ \mathrm{d}h \right)^{-3/5} \tag{6}$$

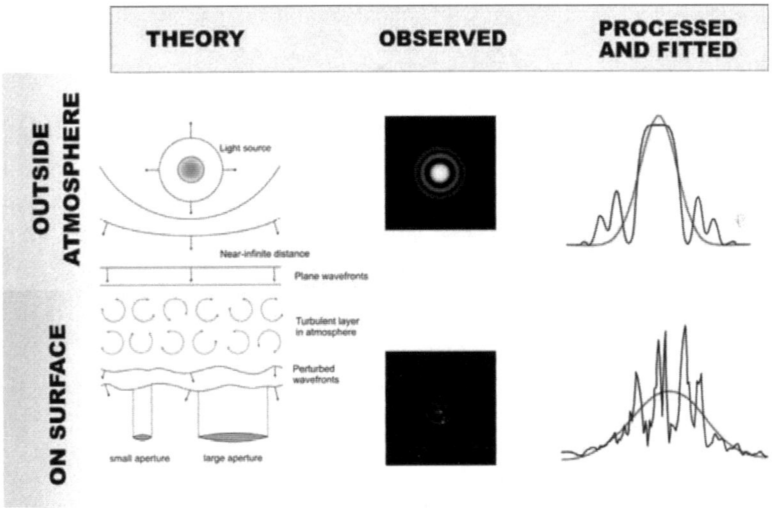

Figure 4. Seeing monitor – technique.

1.2. Combination of techniques and anticipated outcomes

Methods from astronomical seeing and boundary layer meteorology need to be combined to connect optical seeing conditions with atmospheric turbulence.[6] Boundary layer meteorology may be determined by turbulence-resolving numerical modelling with LESNIC[15]. Optical seeing conditions may be determined by implementation of an automated seeing monitor on-site.[2] Seeing monitor data can then be compared with modelled results: (1) to determine whether LESNIC is suitable for modelling seeing conditions; (2) to fine-tune the model to make its seeing quality predictions more accurate. If

IOP Conf. Series: Earth and Environmental Science **13** (2010) 012010 doi:10.1088/1755-1315/13/1/012010

a good correlation between actual seeing quality and the LESNIC model's predicted results can be found, it would be possible to employ meteorological data together with the LESNIC model to select a suitable observing site as well as to forecast seeing quality at the site.
1.3 Some preliminary results using LENIC

2. Conclusion

Seeing is one of the most important characteristics in determining whether a site is suitable for a space geodetic observatory. Sub arc-sec seeing conditions are required for LLR in particular. It would be possible to determine seeing conditions at the Matjiesfontein site by making use of an automated seeing monitor. The LESNIC model will be fine-tuned, using seeing monitor data. This will be used for seeing predictions in future.

References

[1] Combrinck WL, Fourie CJS, Croukamp L and Saunders I 2007 Report on preliminary geotechnical and tropospheric site investigation for a proposed space geodetic observatory near Matjiesfontein in the Great Karoo *South African Journal of Geology* **110** 225-234

[2] Roddier F 1981 The effects of atmospheric turbulence in optical astronomy *Progress in Optics* **19** 281-376

[3] Coulman CE 1985 Fundamental and applied aspects of astronomical seeing *Annual Review of Astronomy and Astrophysics* **23** 19-57

[4] Roggeman MC and Welsh B 1996 *Imaging through turbulence* New York: CRC reference

[5] Erasmus DA 2000 Meteorological conditions affecting observing quality on Mauna Kea *Publications of the Astronomical Society of the Pacific* **98** 254-259

[6] Erasmus DA 1988 and 1996 Relating Seeing Quality To Meteorological conditions: Development of Seeing Quality Forecasts and Improvement of Site Selection Procedures http://www.saao.ac.za/~erasmus/overfsee.htm

[7] Erasmus DA 2000 Meteorological conditions and astronomical observing quality ('seeing') at candidate sites for the Southern African Large Telescope *South African Journal of Science* **96** 1-8

[8] Garratt JR 1992 *The atmospheric boundary layer* Cambridge atmospheric and space science series: Cambridge Univeristy Press 224-227

[9] Kolmogorov AN 1941 The local structure of turbulence in incompressible viscous fluid for very large Reynolds' numbers *Comptes rendus (Doklady) de l'Academie des Sciences de l'URSS* **30** 301-305

[10] Tatarski VI 1961 *Wave propagation in a turbulent medium* New York: Dover Publications 79

[11] Wargau WF 1994 Comparing seeing measurements at SAAO/Sutherland, Gamsberg/Namibia and ESO/La Silla *Monthly Notices of the Astronomical Society of Southern Africa* **53** (9 & 10) 88

[12] Fried DL 1965 Statistics of a geometric representation of wavefront distortion *Journal of the Optical Society of America* **56** (10) 1427-1435

[13] Fried DL 1966 Optical resolution through a randomly inhomogeneous medium for very long and very short exposures *Journal of the Optical Society of America* **56** (10) 1372-1379

[14] Vernin J and Muñoz-Tuñón C 1992 Optical seeing at La Palma Observatory, I. General guidelines and preliminary results at the Nordic Optical Telescope *Astronomy & Astrophysics* **257** 811-816

[15] Esau I 2004 Simulation of Ekman Boundary Layers by Large Eddy Model with Dynamic Mixed Subfilter Closure *Environmental Fluid Mechanics* **4** (3) 273

[16] Esau I 2004 Simulation of Ekman Boundary Layers by Large Eddy Model with Dynamic Mixed Subfilter Closure. *Environmental Fluid Mechanics* **4** (3) 273

[17] Aristidi E, Fossat E, Agabi A, Mékarnia D, Jeanneaux F, Bondoux E, Challita Z, Ziad A, Vernin J and Trinquet H ,2009,Dome C site testing: surface layer, free atmosphere seeing, and isoplanatic angle statistics. *Astronomy & Astrophysics* **499** 955

IOP Conf. Series: Earth and Environmental Science **13** (2010) 012011 doi:10.1088/1755-1315/13/1/012011

A Micro-Meteorological Experiment in the Atmospheric Boundary Layer in Highveld Region

I.N. Esau[1,2], S.S. Zilitinkevich[1,2,3], G. Djolov[4], C.J. deW. Rautenbach[4]

[1] G.C. Rieber Climate Institute of Nansen Environmental and Remote Sensing Center, Thormohlensgate 47, 5006, Bergen, Norway

[2] Bjerknes Centre for Climate Research, Bergen, Norway

[3] Finnish Meteorological Institute, Helsinki, Finland

[4] University of Pretoria, Republic of South Africa

Email: igor.ezau@nersc.no

Abstract. Meteorology of the planetary boundary layer (PBL) is to large extent determined by turbulent processes. Those processes and their interaction with surface properties are not well understood. The processes over heterogeneous land surfaces are understood even less. To progress in the understanding simultaneous observations with a network of meteorological stations are needed. A joint project between Norwegian and South African research foundations funded a micrometeorological experiment in the Highveld area of the South Africa (MMEH). The experiment has been organized to collect data from 5 automatic meteorological stations placed at 7 km to 23 km separation distances from each other. The data were collected continuously over 2 years. This paper presents the idea, the theoretical background and the organization of the MMEH.

1. Introduction

The planetary boundary layer (PBL) is the lowermost layer of the atmosphere where the non-linear processes of the turbulent exchange play an important role. The turbulent exchange is carried out by eddies that are conveniently described by statistical methods (see [1] for review). The statistical approach to the turbulent exchange is justifiable for well developed Kolmogorov's turbulence on scales sufficiently small as compared to the scales of the turbulence generation processes. This is not the case in the PBL. A brief look on the PBL turbulence spectra disclose that the most energetic turbulence is associated with eddies on rather large scales. For clarity and brevity, we will refer here the large eddies in the PBL as vortexes. In a narrow sense the vortexes are associated with the scales of $O(h)$, where h is the PBL depth, which is proportional to the scale of the spectral energy maxima.

The vortexes have been observed not only statistically but also as individual long-living and spatially coherent structures in the PBL. The observational studies have been focused on the marine PBL because the ocean surface maintains the large degree of homogeneity. The surface effects of the vortexes can be easily observed with the synthetic aperture radar as variations of the radar signal brightness on short-living capillary-gravity waves [2–4]. At the same time, the vortex scales are easily identified through cloud structure analysis [5, 6]. An extensive review of observations [7] disclosed that the vortex patterns such as closed or open cloud cells and cloud rolls are ubiquities at all latitudes including tropical and subtropical areas of the South Africa. The cells (see Figure 1) are observed when buoyancy dominates the turbulence production whereas the rolls are observed when the velocity shear dominates.

At the same time, quantitative studies of the vortexes are unusual due to lack of properly sampled spatially distributed data. Recently, several field campaigns such as HATS (The Horizontal Array Turbulence Study) provided data to study the spatial structure of small-scale turbulence. The data can be used to study the vortexes also when the Taylor's hypothesis is applied. However, the largest resolved scales in those experiments were still rather small of about 1 km. In order to collect data for studies of larger vortex scales, a micro-meteorological experiment in the Highveld area of the South Africa (MMEH) was organized in 2008-2010. The paper is aimed to describe this experiment.

Figure 1. Convective clouds over Central-North-East region of the South Africa. Clouds are organized in elongated cloud cells. Clouds prefer greener (darker and presumably wetter) areas even when it requires cell scale modification. Photo is taken by I. Esau from commercial airplane at the height of about 6000 m.

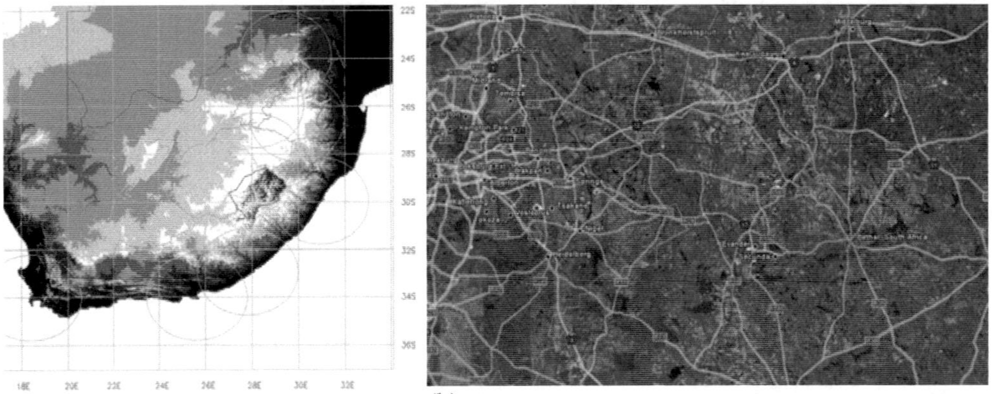

(a) (b)

Figure 2. (a) Topographic map of the South Africa. Shading shows the altitude with the brightest shades corresponding to 1500 m. (b) The map of the Central-North-Eastern region of the South African Republic (after Google Earth). The meteorological station network is placed south of Bethal.

2. Micro-meteorological experiment in the Highveld area (MMEH)

The large part of Republic of South Africa (Figure 2a) is located in dry sub-tropical and tropical areas with predominantly arid and semi-arid climate. Rainfalls are seasonal with the summertime (December through March) maximum and the wintertime (June through September) minimum. Rainfalls are largely convective by nature. The convective clouds are controlled by the PBL turbulence and external parameters such as the atmospheric static stability, surface roughness and the mean wind [8] as well as the moisture availability and therefore the height of the condensation level (LCL). The MMEH rationale suggests that the triggering of the deep convection should be controlled by the soil moisture, scales of the surface heterogeneity and types of vegetation (see Figure 3).

The Highveld region is situated in Central-North-Eastern part of the country (Figure 2b). The area is largely flat with gradual increase of elevation from west to east. At small spatial scales, depressions and hills could be found with the elevation difference of 10-20 m and the typical elevation gradients of 5-10 m km^{-1}. The main economical activities around the experimental site to the north of Bethal town are coal mining, electricity generation and agriculture (maize and wheat). The area is dotted by small water magazines, low albedo pit mines, and townships of 1-2 km in size.

The meteorology of Highveld has been intensively studied over the last 30-40 years. The studies [9] however were focussed more on stably stratified (nocturnal and wintertime) PBL than on daytime convection. It was motivated by the air quality monitoring needs [10]. Tennant and Hewitson [11] published a comprehensive analysis of the rainfall climatology at 276 stations. The climatology encompasses 1936 to 1999. It is based on a self-organized map method (a kind of hierarchical cluster analysis) to distinct homogeneous sub-regions. Highveld has been put into the North-Eastern Interior homogeneous rainfall region. The average number of days with precipitations > 1 mm is between 30 and 40. The average seasonal amount of precipitations is 350 mm to 450 mm. The 50 year climatology analysis in this region indicates increasing trends of the number of rain days while the spatial picture of trends is mosaic [12].

Thomas et al. [12] study put forward a problem, which is at the edge of physical and social climatology, that could be addressed only with research studies of the MMEH type or similar. The study showed that the perception of the climate change impact does not correlate significantly with the measured social impact. For instance in the region south of the Highveld area, 87% of the respondents identified the climate change as the observed challenge or risk while no measurable changes has been found in the region's climate by 1999. In Highveld, the observed changes are more pronounced. So, farmers have already initiated practical measures to mitigate the changes in the temporal and spatial patterns of the moisture availability. But the unguided mitigation process is chaotic, based rather on try-and-fail approach. While serving for improvement for some farmers at short time scales, it may lead to bigger challenges at long scales. In this respect, the project "Analysis and Possibility for Control of Atmospheric Boundary Layer Processes to Facilitate Adaptation to Environmental Changes" in the frameworks of the South African – Norwegian Programme for Research and Co-operation Phase II and the MMEH experiment are attempts to create some theoretical and observational background for the scientifically guided mitigation process.

Table 1. List of automatic meteorological stations, their coordinates and altitude

Station and farm name	Southern latitude	Eastern longitude	Altitude above sea level
S1. Jan de Jager, Banklaagte	26.40530	29.56912	1650
S2. Anton van Tonder, Yzervarkfontein	26.37059	29.45499	1660
S3. Bram Jordan, Rietkuil	26.28638	29.61690	1670
S4. Anton Pelse, Driefontein	26.08918	29.56606	1706
S5. Daleen von Wieligh, Bultfontein	26.12696	29.49994	1656

IOP Conf. Series: Earth and Environmental Science **13** (2010) 012011 doi:10.1088/1755-1315/13/1/012011

The area south of Bethal town was chosen as the meteorological test-bed for data collection. The area is geographically and economically homogeneous and thought to be representative for the whole Highveld region. Figure 4 shows the locations of the automatic meteorological station network in this area. In total, 5 stations were placed on farmers' land, mostly at the edge of agricultural (maize) fields. The station placing is not ideal because practical security reasons and logistics have to be taken into account. Some stations are located close to some low buildings, farmer houses and wire fences, which may contaminate the collected data. The complete list of stations with their coordinates and altitudes is in Table 1.

The experimental site is a flat, slightly inclined plain at altitudes from 1650 m to 1706 m. Thus the total elevation gradient is just 56 m with no elevations lower than 1630 m and higher than 1715 m between the stations. The area is rather densely populated but the population is living in scattered low houses often surrounded by a group of 10-15 m trees. The rest of the area is heterogeneous agricultural, mostly maize, fields. Thus the surface heterogeneity is generally small and dependent on soil moisture. In larger area there are also open coal mines that have much lower albedo. Figure 5 presents a typical view at the station site.

The stations are placed in an approximate cross pattern forming two groups. The southern group S1, S2 and S3 forms a triangle with sides 12 km (S1 to S2), 19 km (S2 to S3) and 14 km (S1 to S3). The northern group consists of two stations S4 and S5 placed 30-35 km north from the southern group and separated by 8 km. Thus the experiment site pattern allows studies of the turbulence coherency on scales from 8 km (S4 to S5) to 35 km (S1 to S4). The distances between all stations are presented in Table 2.

Table 2. Distances between meteorological stations given in km.

	S2	S3	S4	S5
S1	12.006	14.054	35.152	31.709
S2		18.656	33.195	27.459
S3			22.507	21.222
S4				7.825

The data is sampled continuously. The samples are averaged automatically by meteorological stations over intervals 10-30 min depending on the concrete station set up and the period of observations. The averaged values from temperature, humidity and wind sensors are stored into station's digital loggers. The data collection started in March 2008. Since the stations are placed in a relatively remote place, frequent regular access would have overrun the project budget. Therefore, the calibration, repair and general check of the stations' performance are carried out about once a month. It results in data collection irregularities and rather large data gaps. Table 3 gives the preliminary data recovery rate by months through September 2008 to January 2009. To study the dependence on the averaging and reduce the logger memory requirements, the data averaging interval was set to 30 min at some stations in September-October and 10 min in November-January.

The complementary data on the rainfall and cloudiness are obtained from meteorological radars in Ermelo (to the east) and Irene (to the west). The processed data [13] and combined with gauge and satellite data radar maps for daily averaged precipitations with resolution up to 1 km are available on-line from (http://metsys.weathersa.co.za/radar_image/precip.24hr/archive/). The procedure has been developed in the project SIMAR. The radars operate in a volume-scan mode with ~5 minute time resolution. The data processing system TITAN is used to manipulate and display radar data and the mdv-format is used for all data and products (all originated from NCAR, USA).

IOP Conf. Series: Earth and Environmental Science **13** (2010) 012011 doi:10.1088/1755-1315/13/1/012011

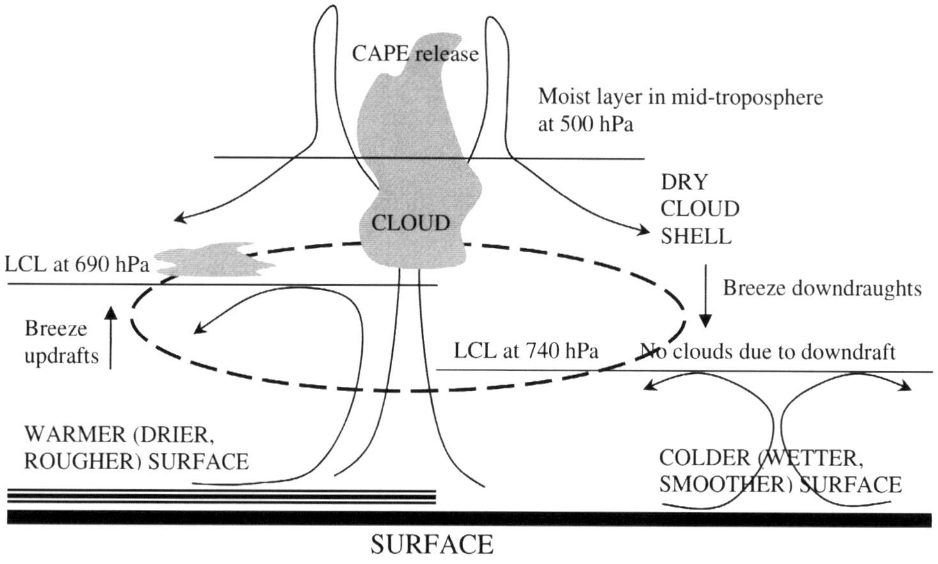

Figure 3. The scheme of the turbulent vortex circulation expected to be observed over heterogeneous surface as effect of dynamics-thermodynamics interactions and latent to sensible heat flux interchange.

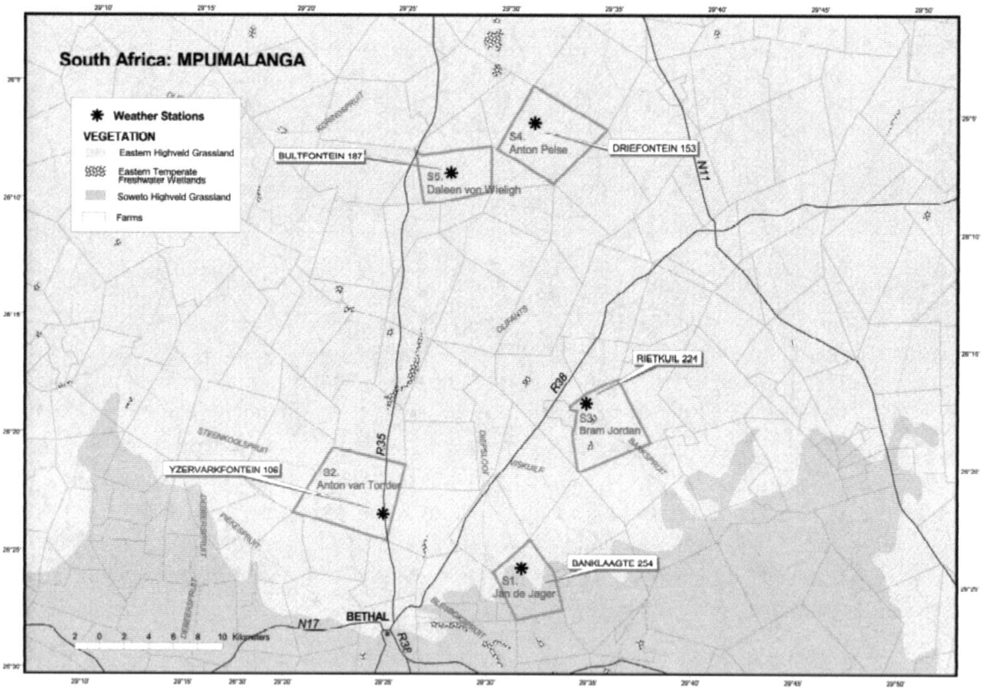

Figure 4. Location of the automatic meteorological stations in the experiment. Stars denote the weather stations; gray contours are farm boundaries; blue lines are significant roads. Farm names are given in red and farm addresses are within yellow boxes.

Figure 5. A typical view at the weather station site. Site S1 on Jan de Jager's farm.

Table 3. Data completeness (in % of the total possible coverage by time) at meteorological stations as achieved through September 2008 to January 2009. Observe that not all logged data are retrieved in the database yet. Such data denoted as NP – "not processed".

Station	Sep	Oct	Nov	Dec	Jan
S1.	60	100	97	54	58
S2.	60	100	58	32	58
S3.	58	NP	79	60	NP
S4.	NP	NP	NP	NP	NP
S5.	60	NP	37	NP	NP

3. Davis Vantage Pro 2 automatic meteorological station

Networks of wireless weather stations gradually become essential for investigation of the atmospheric processes over strongly heterogeneous surfaces. For example, Nadeau et al. [14] deployed 92 weather stations Sensorscope in the central part of a university campus covering an area of 300 by 400 m. The data (temperature, wind speed and direction) where sampled at the rate of 120 s with the averaging over 30-min periods. This high temporal and special resolution meteorological measurements were used to evaluate the applicability of Monin-Obukhov similarity theory for the urban surface fluxes.

Davis Vantage Pro 2 is an automatic meteorological station with independent energy supply from solar panels and wireless data transmission from a set of configurable meteorological sensors to autonomous storage and display module. This equipment has several strong advantages. It is relatively inexpensive on the price per station of about $3000. So the modest project budget could be used to purchase several or even several tens of stations. The stations are automatic. They can operate, collect, log and store data without manual maintenance, without external source of the energy supply and without frequent access to sensors and data. All these features are not only reducing the cost but also make it possible to place stations in areas with little or no infrastructure. The wireless transmission of data with the frequency hopping spread spectrum radio is able to support high transmission rate up to 1000 Hz. It does not require application for a special permit and works robustly on the range up to 100 m. More information about the station and its installation could be found in (Application Note 30).

Each of the stations are equipped with a pressure sensor, 2 temperature sensors, humidity sensors, wind anemometers, rain collection gauges and radiation measurements. This equipment allows for gradient measurements and calculation of the complete surface radiation balance and turbulent fluxes.

An important issue of the experimental meteorology is the quality characteristics of data and stability of the sensors' readings. Calibration and quality of measurements have been assessed independently [15]. He concluded that temperature sensors demonstrated good behaviour with the scaling factor ~1.0 and negligible offset. Humidity sensors demonstrated acceptable behaviour with the scaling factor above 0.9 and the offset ~6 %. However, the humidity values during wet conditions were too high. The pressure sensor has adjusted for real air pressure (not barometric pressure reduced to sea level) after several hours of working. Its behaviour was also good with the scaling factor 0.97. Wind anemometers were working acceptably good with the scaling factor 0.89 and negligible offset. Solar radiation sensors were tested against ventilated CM21 Kipp & Zonen pyranometer. Uncorrected Vantage Pro radiation data were lower by about 7% with scaling factor about 1.07 and a small offset. But the data scattering was non-negligible so longer time of averaging is required to obtain a reliable radiation flux. Applying the manufacturer's temperature coefficient of 0.12% per °C (above 25°C) improved the scaling factor to 1.04. In short, this comparison shows that the Vantage Pro Plus delivers excellent to good quality data, compared to professional-grade sensors.

The overall conclusion [15] can be formulated as follows. Certainly the Davis Automatic Weather Station accuracy is way too low to detect minor long-time trends in meteorological parameters. This conclusion opposes the statement of the producer on the station abilities. Thus, the equipment is not suitable for representative meteorological observations, at least without frequent calibration. Nevertheless, the equipment can be used for short-time micro-meteorological observations, especially those like the MMEH, which rely on statistically significant features but not on the individual readings. The data format is not carefully designed. A lot of important information is missed or stored improperly. This is especially related to the timing and sensor's location of readings. At the same time, the format reserves bytes for many derived parameters that can be computed by request. It creates significant complications for the data processing as well as reduces the amount of data to be stored in the limited logger memory.

The Davis Vantage Pro equipment has been already used in many educational and scientific projects, e.g. in the Citizen Weather Observer Program (CWOP). In Africa, the equipment has been used in BodEx-2005 (Bodele (in Chad) Experiment) meteorological experiment [16, 17] aimed to quantify the dust aerosol production. In this experiment, the focus was on measurements of the near-surface winds and their correlations to the aerosol load in the lower atmosphere. As the publications disclose the choice of the Davis Vantage Pro equipment was scientifically justified and paid of with research data of high quality.

5. Conclusions

Our experience has shown that it is feasible to organize relatively inexpensive micro-meteorological experiment based on affordable mass-production meteorological equipment, automated data collection, logging and storage, and utilization of complementary information. The experience has also highlighted several difficulties that need to be taken seriously at the stage of the experiment planning. Firstly, the experiment should be well thought of in terms of the equipment safety and accessibility. Secondly, the data should go through pre-processing as often as possible in order to find irregularities in the equipment operation and calibration. The routine has not been implemented in the MMEH, which caused somewhat lower data recovery rate. Thirdly, the accessibility of the complementary meteorological information has to be secured in advance.

References

[1] Zeman, O., 1981. Progress in the Modeling of Planetary Boundary Layers, *Ann. Rev. Fluid Mech.*, **13**, 253-72

[2] Mourad, P. D., 1996. Inferring the multiscale structure of atmospheric turbulence using satellite-based SAR imagery, *J. Geophys. Res.*, **101**(C8) 18433-18499

[3] Savtchenko, S., 1999. Effect of large eddies on atmospheric surface layer turbulence and the underlying wave field, *J. Geophys. Res.*, **104**, 3149–3157

[4] Levy, G., 2001. Boundary layer roll statistics from SAR, *Geophys. Res. Lett.*, **28**, 1993-1995

[5] Agee, E.M., 1984. Observations from space and thermal convection: A historical perspective, *BAMS*, **65**, 938-949

[6] Agee, E.M., 1987. Mesoscale cellular convection over the oceans, *Dyn. Atmos. Oceans,* **10,** 317–341

[7] Atkinson, B.W., and J., Wu Zang, 1996. Mesoscale Shallow Convection in the Atmosphere, *Rev. Geophys.*, **4**, 403–431

[8] Zilitinkevich, S.S., J.C.R. Hunt, A.A. Grachev, I., Esau, D.P. Lalas, E. Akylas, M. Tombrou, C.W. Fairall, H.J.S. Fernando, A. Baklanov, and S.M. Joffre, 2006. The influence of large convective eddies on the surface layer turbulence, *Quart. J. Roy. Meteorol. Soc.*, **132**, 1423-1456

[9] Jury, M.R., and G.R. Tosen, 1989. Characteristics of the winter boundary layer over the African Plateau: 26 ° S, *Boundary-Layer Meteorol*, **49**(1-2), 53-76

[10] Scheifinger, H., and G. Held, 1997. Aerosol behaviour on the South African Highveld, *Atmos. Environ.*, **31**(21), 3497-3509

[11] Tennant, W.J., and B.C. Hewitson, 2002. Intra-Seasonal Rainfall Characteristics and their Importance to the Seasonal Prediction Problem, *Int. J. Climatol.*, **22**, 1033-1048

[12] Thomas, D.S.G., Twyman, C., Osbahr, H., and B.C. Hewitson, 2007. Adaptation to climate change and variability: farmer responses to intra-seasonal precipitation trends in South Africa, *Climatic Change*, **83**, 301–322

[13] Pegram, G.G.S., A.W. Seed, and D.S. Sinclair, 2002. Comparison of Methods of Short-Term Rainfield Nowcasting. *27th EGS Assembly*, Nice, France. April.

[14] Nadeau, D.F., Brutsaert, W., Parlange, M. B., Bou-Zeid, E., Barrenetxea, G., Couach, O., Boldi, M.-O., Selker, J.S. and M. Vetterli, 2009. Estimation of urban sensible and heat flux using a dense wireless network of observations, Envron. Fluid Mech., 9, 635-653

[15] Massen, F., 2003. A short calibration study of the Vantage Pro Plus weather station, (last accessed 11 March 2009) available on http://meteo.lcd.lu/papers/comparison_vantage/vantage_calibration.html

[16] Giles, J., 2005. Climate science: The dustiest place on Earth, *Nature*, **434**, 816-819

[17] Washington, R., Todd, M.C., Engelstaedter, S., Mbainayel, S., and F. Mitchell, 2005: Dust and the Low Level Circulation over the Bodélé Depression, Chad: Observations from BoDEx 2005, *J. Geophys. Res.*, **111**(D3), D03201.

IOP Conf. Series: Earth and Environmental Science **13** (2010) 012012 doi:10.1088/1755-1315/13/1/012012

A Study of Stable Atmospheric Boundary Layer over Highveld South Africa

P. Luhunga[1], I. Esau[2,3], G. Djolov[1]

[1] University of Pretoria, Republic of South Africa

[2] G.C. Rieber Climate Institute of Nansen Environmental and Remote Sensing Center, Thormohlensgate 47, 5006, Bergen, Norway

[3] Bjerknes Centre for Climate Research, Bergen, Norway

Email: george.djolov@up.ac.za

Abstract

The study is part of the South African – Norwegian Programme for Research and Co-operation Phase II "Analysis and Possibility for Control of Atmospheric Boundary Layer Processes to Facilitate Adaptation to Environmental Changes". The research strategy of the project is based on 4 legged approach. 1) Application and further development of contemporary atmospheric boundary layer theory. 2) Use of modeling based on large eddy simulation techniques. 3) Experimental investigation of turbulent fluxes. 4) Training and developing academics capable of dealing with the present and new challenges. The paper presents some preliminary results on the micrometeorological variability of the basic meteorological parameters and turbulent fluxes.

1.Introduction

Atmospheric boundary layer (ABL) is the part of troposphere that is directly influenced by the presence of the earth's surface and responds to surface forcing with a time scale of about one hour or less. It plays an important role in the dynamic state of the whole atmosphere by controlling the exchange of heat, moisture, momentum, aerosols, gasses, pollutants and trace gases relevant to the climate change phenomenon. The success of daily weather prediction, climate simulations, air pollution and urban meteorology studies, aviation and agriculture industry are crucially dependent on the proper parameterization of the ABL dynamics.

The stable boundary layer (SBL) study over the Highveld South Africa has a special relevance since the majority of the electric power generating plants is located in this region. SBL is characterized by light (calm) wind near the surface, virtual potential temperature increase with height, humidity decrease with height, strong wind shears (directional & speed), turbulence generated mechanically by shear is destroyed by negative buoyancy and viscosity. The non-turbulent periods are frequent and SBL becomes decoupled from the surface. The oscillatory type of behavior of the near-surface atmospheric variables is a manifestation of the non-linear character of the turbulent exchange.

The aim of the paper is to study the SBL characteristics over the Highveld region of South Africa using automatic observational network data and theoretical models. In particular some preliminary results of the time variation of meteorological parameters and turbulent fluxes of momentum and heat are presented.

2. Data and calculation of the turbulent fluxes

The experimental set up for the observation of all relevant meteorological variables and the data collection are presented in paper [1], which is part of the present international conference publication. It should be noted that in addition to the Davis station experimental set up, which was mounted at height 2 m, one more temperature sensor has been installed at 0.5 m to allow for the calculation of turbulent fluxes. The calculation of the turbulent fluxes is done using the collected experimental data during 2008 and 2009 observational campaigns. The subset of SBL condition data was organized using the temperature difference between 2m and 0.5m.

The determination of the turbulent fluxes of momentum and heat is based on the recent advancement of the similarity theory method. First, for the past already 60 years Kolmogorov's [2] approach for turbulent closure models based on the turbulent kinetic energy (TKE) balance has been a major scientific tool. His hypothesis, however, is theoretically justified for-neutrally stable turbulent flows only. Many attempts to apply Kolmogorov's method for stratified flows have Ri_{cr} encountered difficulties. The straightforward application of the TKE budget equation leads to the existence of critical Richardson number. The experimental data from laboratory and atmospheric flows have persistently shown that turbulence exists at higher than the critical Richardson number values.

Recently (see [3]) it has been shown that the use of the total turbulent energy TTE=TKE +TPE (TPE is the turbulent potential energy), instead of TKE only, is a promising way for development of consistent turbulent closure models. One main result of the new theory is that the critical Richardson number does not separate the turbulent and the laminar flow regimes. Instead turbulence exists at any Richardson number.

Secondly, the Monin –Obuhkov (MO) [6] similarity theory, which postulated that the vertical variation of mean flow and turbulence characteristics in the surface layer should depend only on the surface momentum flux as measured by friction velocity u_*, the buoyancy flux F_θ, and the height z, can not be considered as the ultimate framework for studying the surface layer turbulence. According to MO the Richardson number $Ri \equiv \beta(d\Theta / dz)/(dU / dz)^2$, monotonically increases with z / L, and at $z / L \to \infty$ has an asymptote maximum value of $Ri_{cr} = \chi^2 C_{\Theta 1} k_T^{-1} C_{U1}^{-2}$. This shows that the similarity theory universal function range can not exceed $Ri > Ri_{cr}$. This is a classical result which follows from the equation of kinetic turbulent energy balance accepted at the time when MO theory was formulated. However, as it has been shown in [3] turbulence is not fully suppressed at any Richardson number.

In a papers [4], [5] the classical MO theory has been generalized by considering the fact that the boundary layer turbulence depends on the free atmosphere parameters. In addition to the Monin-Obukhov length L, which characterize the effect of local buoyancy on turbulence, there are two additional length scales: L_f - describing the effect of the

Earth rotation and L_N - characterizing the non-local effect of the static stability of free atmosphere. The definition of these parameters is:

$$L_f = \frac{\tau^{1/2}}{|f|}$$

$$L_N = \frac{\tau^{1/2}}{N}$$

where f is the Coriolis parameter, and $N = (\beta\partial\Theta/\partial z)^{1/2}$ is the free atmosphere Brunt-Vaisala frequency. Interpolation between the squared reciprocals define a new composite turbulence length scale

$$\frac{1}{L_*} = \left[\left(\frac{1}{L}\right)^2 + \left(\frac{C_N}{L_N}\right)^2 + \left(\frac{C_f}{L_f}\right)^2\right]^{1/2}$$

where $C_N = 0.1$ and $C_f = 1$ are empirically determined dimensionless constants.

In this paper the momentum and heat resistance laws are solved using the widely accepted functional form for the universal functions for stable stratification conditions [7].

$$\varphi_m(\frac{z}{L}) = 1 + 4.7\frac{z}{L}$$

$$\varphi_\theta(\frac{z}{L}) = 0.74 + 4.7\frac{z}{L}$$

The linear form of the above equations allows for obtaining analytical solutions for the turbulent fluxes.

3. Micrometeorological variability of basic turbulence characteristics

The data in section 2 and the generalized theory discussed in the previous section allow presentation of some preliminary results for the variation of the meteorological variables and turbulent fluxes in stable stratification conditions over the Highveld region. In Figure 1 the monthly distribution of the rainfall and rainfall intensity are presented. It should be immediately noted the substantial difference between these parameters for stations, which are not too far apart. The analysis of individual cases for rainfall together with the turbulent fluxes will allow the turbulent vortex circulation expected over heterogeneous surface as an effect of dynamics-thermodynamics interactions and latent to sensible heat flux interchange.

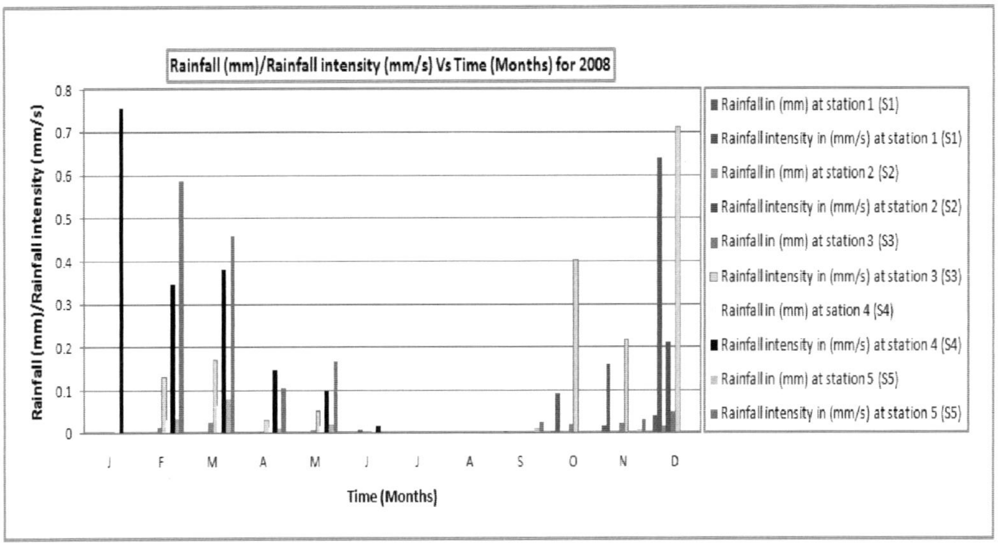

Figure 1. Rainfall and rainfall intensity distribution.

The extreme variability of the stability conditions at micrometeorological scales can be also observed in Figure 2, which presents the Monin-Obukhov length variation for a particular night.

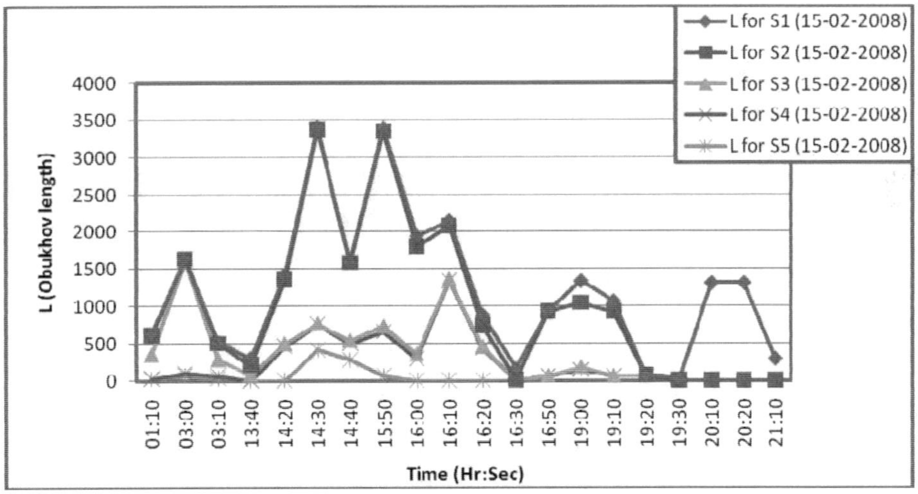

Figure 2. Monin-Obukhov length variation

Figures 4 and Figure 5 demonstrate the big difference between the turbulent fluxes at micrometeorological scale for the same particular night period as in Figure 2.

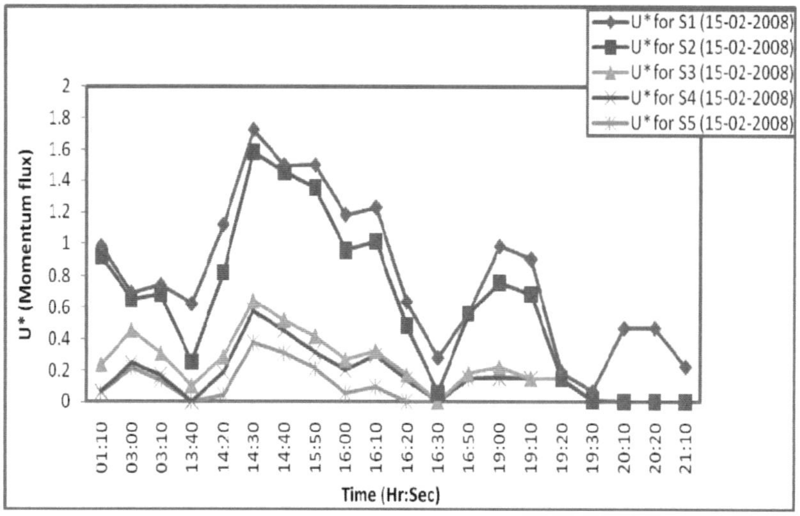

Figure 4. Variation of turbulent friction velocity.

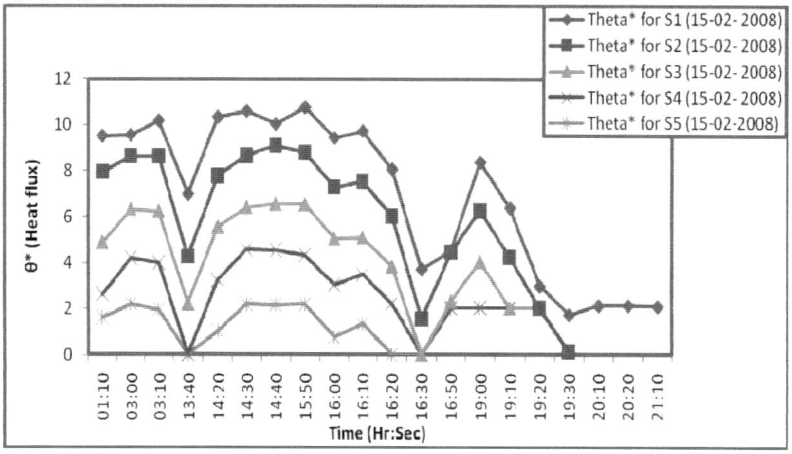

Figure 5. Variation of the heat flux.

4. Conclusions

The preliminary results presented in the paper reveal the essential variability of the SBL characteristics over Highveld region, South Africa. They are of importance for the solution of the heavy air pollution episodes in the region and long-range pollutants transports to the neighboring countries since it contains most of the coal power generating stations of South Africa. The strong variations of the micro-scale rain patterns phenomenon needs also of properly sampled spatially distributed data. Finally, the second phase of the research will concentrate on the mechanisms of convective clouds development which are controlled by the PBL turbulence and external parameters such as the atmospheric static stability, surface roughness and the mean wind as well as the moisture availability and therefore the height of the condensation level.

References

[1] Esau, I.N., Zilitinkevich, S.S., Djolov, G.D and J. deW. C. Rautenbach, 2010: A micro-meteorological experiment in the atmospheric boundary layer in Highveld region, *Paper in the present volume.*

[2] Kolmogorov AN, 1941: Energy dissipation in locally isotropic turbulence, *Doklady AN SSSR,* **32**(1):19-21.

[3] Zilitinkevich SS, Elperin T, Kleorin N and I Rogachevski, 2007, Energy and flux budge (EFB) turbulent closure model for stably flows, Part I: steady-state, homogeneous regimes., *Atmospheric Boundary Layers, Springler,*11-36.

[4] Zilitinkevich S and I.N. Esau, 2005, Resistance and heat transfer laws for stable and neutral planetary boundary layers: old theory, advanced and reevaluated, *Quart J Roy Meteorol Soc.* , **131**, 3339-3356.

[5] Zilitinkevich SS and I.N.Esau, 2007, Similarity theory and calculation of turbulent fluxes at the surface for the stably stratified atmospheric boundary layer, *Atmospheric Boundary Layers, Springer*, 37-50.

[6] Monin AS and AM Obukhov, 1954, Main characteristics of the turbulent mixing in the atmospheric surface layer, Trudy Geophys Inst AN SSSR, **24**(151), 153-187.

[7] Businger JA, Wyngaard JC, Uzumi Y and EF Bradley, 1971, Flux-profile relationship in the atmospheric surface layer, *J Atm Sci*, **28**, 181-189.